AutoUni – Schriftenreihe

Band 41

AutoUni – Schriftenreihe

Band 41

Herausgegeben von
Volkswagen Aktiengesellschaft
AutoUni

Jiří Jerhot

Environmental perception with self-diagnosis for advanced driver assistance systems

Integrated probabilistic approach

Logos Verlag Berlin

AutoUni – Schriftenreihe

Herausgegeben von
Volkswagen Aktiengesellschaft
AutoUni
Brieffach 1231
38436 Wolfsburg
Tel.: +49 5361-896-2207
Fax: +49 5361-896-2009
http://www.autouni.de

Bibliografische Information der Deutschen Nationalbibliothek

Die Deutsche Nationalbibliothek verzeichnet diese Publikation in der Deutschen Nationalbibliografie; detaillierte bibliografische Daten sind im Internet "uber http://dnb.d-nb.de abrufbar.

Zugleich: Dissertation, TU Braunschweig, 2013

ISBN 978-3-8325-3348-9
ISSN 1867-3635

Logos Verlag Berlin GmbH
Comeniushof, Gubener Str. 47,
10243 Berlin
Tel.: +49 30-42 85 10-90
Fax: +49 30-42 85 10-92
http://www.logos-verlag.de

Technische Universität Carolo-Wilhelmina zu Braunschweig

Environmental perception with self-diagnosis for advanced driver assistance systems

Von der Fakultät für Elektrotechnik, Informationstechnik, Physik
der Technischen Universität Carolo-Wilhelmina zu Braunschweig
zur Erlangung der Würde eines Doktor-Ingenieurs (Dr.-Ing.)
genehmigte Dissertation von

Jiří Jerhot
aus Prag

eingereicht am: 02. Juli 2012
mündliche Prüfung am: 16. Januar 2013

Referenten: Prof. Dr.-Ing. W. Schumacher
Prof. Dr.-Ing. B. Michaelis
Prof. Dr.-Ing. T. Form

2013

Acknowledgments

This thesis results from my work at the Research Division of Volkswagen AG in Wolfsburg. It was developed within the project "Integrated lateral assistance", a subproject of research initiative "AKTIV" supported by the German Federal Ministry of Economics and Technology.

First of all I am very grateful to my supervisor Prof. Thomas Form for his long-time kind support and all the constructive discussions and helpful hints. My further thanks go to Prof. Walter Schumacher and Prof. Bernd Michaelis for co-supervising this thesis and their assistance in the final phase.

My special thanks go to Dr. Marc-Michael Meinecke for his continuous encouragement, motivation and intensive support during this work.

I would like to thank Dr. Jan Effertz and my later office neighbors Dr. Jörn Knaup and Dr. Thien-Nghia Nguyen for fruitful discussions and pleasant working environment in the final phase.

I am also grateful to Dr. Holger Philipps for his support in the initial phase of this work, and Dr. Alexander Kirchner and further colleagues from the former sub-department of Driver Assistance Electronics, especially Dr. Bernd Rössler, Dr. Marian-Andrzej Obojski, Dr. Christian Brenneke, Dr. Dirk Stüker, Markus Köchy and Dr. Christian Koelen for the always friendly working atmosphere.

Furthermore, I would like to thank my colleagues from Audi AG Andreas Siegel and Dr. Richard Altendorfer and my former students Deniz Benli and Michael Meinecke for various technical support.

Finally, I would like to express my gratitude to my parents who have aways believed in me, my brother Filip who proofread my English draft and my wife Elena who supported me with love and patience during my work on this thesis.

Abstract

Today, advanced driver assistance systems (ADAS) represent an increasing contribution to active road safety and driving comfort. Their task is to help the driver avoiding accidents or at least to minimize their consequences. Depending on their objectives they can support the driver's decisions by providing additional information (e.g. Park Distance Control, Night Vision, Traffic Sign Recognition) or even directly influence the driving process (e.g. Park Assist, ACC, Lane Assist). Since the absolute number of ADAS sensors in vehicles is permanently increasing, further development of existing sensor data processing mechanisms is required to ensure a robust functionality and (eventual) timely detection of system limits (e.g. caused by sensor misalignment, bad weather conditions, pollution or aging). Therefore, continuous knowledge of the quality of the environmental perception is of significant importance.

In this thesis, general probabilistic approach to multi-sensorial environmental perception of ADAS is presented. This approach incorporates sensor data fusion with self-diagnosis capability and maneuver level intent estimation of detected objects. Thus, the quality of environmental perception can be continuously monitored and the intents of the traffic participants can be predicted. The resulting probabilities are uniform and consistent basis and reflect the reliability of the results. This knowledge is an important prerequisite for the development of future complex and robust driver assistance systems.

The developed concepts have been used and approved in project "Integrated Lateral Assistance", a subproject of research initiative AKTIV (abbreviation for "Adaptive and Cooperative Technologies for the Intelligent Traffic") supported by the German Federal Ministry of Economics and Technology.

Kurzfassung

Heutige Fahrerassistenzsysteme leisten einen zunehmenden Beitrag zur aktiven Sicherheit im Straßenverkehr und zur Erhöhung des Fahrkomforts. Ihre Aufgabe ist es den Fahrer bei der Vermeidung von Kollisionen zu unterstützen oder zumindest deren Folgen zu minimieren. Je nach Ausprägung können sie die Entscheidungen des Fahrers durch Bereitstellung zusätzlicher Informationen unterstützen (z.B. akustische Einparkhilfe, Nachtsicht-Anzeige, Verkehrszeichenerkennung) oder sogar direkt das Fahrverhalten beeinflüssen (z.B. Park Assist, ACC, Lane Assist). Da die Gesamtzahl der Fahrerassistenz-Sensoren in Fahrzeugen ständig zunimmt, ist eine Weiterentwicklung der bestehenden Sensordatenverarbeitungsmechanismen erforderlich, die eine robuste Funktionalität und die (eventuelle) rechtzeitige Erkennung von Systemgrenzen (z.B. durch einen dekalibrierten Sensor, schlechte Wetterbedingungen, Verschmutzung oder Alterung) sicherstellen. Daher ist eine kontinuierliche Überwachung der Qualität der Umfeldwahrnehmung von großer Bedeutung.

In dieser Arbeit wird ein allgemeiner probabilistischer Ansatz zur multisensorischen Umfeldwahrnehmung von Fahrerassistenzsystemen vorgestellt. Dieser Ansatz beinhaltet Sensordatenfusion mit einer Selbstdiagnosefähigkeit und Absichtserkennung der detektierten Objekte auf Manöverebene. So kann die Qualität der Umfeldwahrnehmung kontinuierlich überwacht und die Absichten der Verkehrsteilnehmer vorausgesagt werden. Die resultierenden Wahrscheinlichkeiten bilden eine einheitliche und konsistente Grundlage und reflektieren damit die Vertrauenswürdigkeit der Ergebnisse. Diese Kenntnis ist eine wichtige Voraussetzung für die Entwicklung von zukünftigen komplexen und gleichzeitig robusten Fahrerassistenzsystemen.

Die erarbeiteten Konzepte wurden in einem Teilprojekt der Forschungsinitiative AKTIV (Adaptive und Kooperative Technologien für den Intelligenten Verkehr) namens "Integrierte Querführung" eingesetzt. Diese Forschungsinitiative wurde von den deutschen Bundesministerien für Wirtschaft und Technologie gefördert.

Contents

Abbreviations

ABS	Anti-lock Braking System
ACC	Adaptive Cruise Control
ADAS	Advanced Driver Assistance System(s)
ADTF	Automotive Data and Time-Triggered Framework
AKTIV	Adaptive and Cooperative Technologies for Intelligent Traffic
BAS	Brake Assist System
BN	Bayesian Network
CAN	Controller Area Network
CARSENSE	Sensing of Car Environment at Low Speed Driving
COM	Component Object Model
CPT	Conditional Probability Table
DAS	Driver Assistance System(s)
DBN	Dynamic Bayesian Network
DGNSS	Differential Global Navigation Satellite System
DGPS	Differential Global Positioning System
DTC	Diagnostic Trouble Code
EKF	Extended Kalman Filter
EM	Environmental Model
EPS	Electromechanical Power Steering
ESP	Electronic Stabilization Program
FMCW	Frequency Modulated Continuous Wave
FMSK	Frequency Modulated Shift Keying
FN	False Negative(s)
FP	False Positive(s)
GNSS	Global Navigation Satellite System
GPS	Global Positioning System
HC	Heading Control

HMI	Human-Machine Interface
HMM	Hidden Markov Model
INVENT	Intelligent Traffic and User-oriented Technology
KF	Kalman Filter
LDW	Lane Departure Warning
MF	Measurement Fusion
NV	Night Vision
OBD	On-Board Diagnostics
PA	Park Assist
PDC	Park Distance Control
PDF	Probability Density Function
PReVENT	Preventive Safety
QoS	Quality of Service
SDF	Sensor Data Fusion
SMILE	Structural Modeling, Inference, and Learning Engine
TLC	Time to Lane Change
TSR	Traffic Sign Recognition
TTC	Time To Collision
USB	Universal Serial Bus

Chapter 1

Introduction

The vehicle density on roads worldwide has been continuously increasing in the last decades and this trend will likely be kept. In contrast, the factor of "safety" becomes a more and more important role both from the technological and from the political point of view. The resulting dilemma demands various measures such as stepping up checks on road traffic, deploying new road safety technologies, improving road infrastructure and measures to improve users' behavior.

Vehicle manufacturers have introduced a number of safety and assistance systems in the recent years. The aim of such systems is to make the driving safer and more comfortable as well. The reduced work load of the driver decreases the accident risk especially in potentially critical situations. The task of such systems is to help the driver to avoid accidents or at least to minimize their consequences. Depending on their objectives they can support the driver's decisions by providing additional information (e.g. Park Distance Control, Night Vision, Traffic Sign Recognition) or even directly intervene in the driving process (e.g. Park Assist, Adaptive Cruise Control, Lane Assist).

The increasing number of such on-board assistance systems and their growing degree of support require new measures to deal with the complexity and ensure robustness.

1.1 Background

The concepts introduced in this thesis have been used and approved in project "Integrated Lateral Assistance", a subproject of German research initiative "AKTIV" (abbreviation for "Adaptive and Cooperative Technologies for the Intelligent Traffic") supported by the Federal Ministry of Economics and Technology (s. section 2.1.2). AKTIV was the logical continuation of the research project INVENT which was funded by the German Federal Ministry of Education and Research (s. section 2.1.2). The two main projects of AKTIV research initiative were "Active Safety" and "Traffic Management" with numerous subprojects.

The general objectives of AKTIV were:

- Increasing of active safety for vehicles and traffic management systems.
- Decrease of driver load.
- Harmonizing of traffic flows.

The subproject "Integrated Lateral Assistance" was integrated in the "Active Safety" project and it took place between September 2006 and August 2010. The objective of this subproject was to develop a driver assistance system with continuous, integrated lateral and longitudinal guidance in the full speed range from 0 to 180 km/h. This system should support the driver adaptively in lane-keeping and cruise controlling on highways and well built roads under consideration of other traffic participants (incl. static boundary objects). The project focused especially on robustness and reliability of the developed system even in complex situations like e.g. inside of construction sites.

1.2 Motivation

Advanced driver assistance systems (ADAS) support the driver in his routine tasks. The resulting reduced load on the driver decreases the accident risk.

Today's systems are usually based on a single sensor. Thereby, the complete processing chain from sensor measurements to the according command on the actor and/or human machine interface is processed by a single controller unit. The synergy effect potential between multiple assistance systems remains widely unused.

The increasing number of such systems in the vehicles as a consequence of advanced safety measures realized by vehicle manufacturers increases the entire system sophistication and variability. It is obvious, that new diagnostic methods become necessary to deal with the resulting system complexity.

1.3 Objectives

Precise knowledge of the vehicle environment constitutes an essential prerequisite for a driver assistance system to act appropriately in the specified spectrum of traffic situations.

The general aim of the "Integrated Lateral Assistance" project was to develop a system assisting the driver in certain tasks as introduced in section 1.1. This thesis focuses on its environmental perception part. The global objective is to develop a robust modular approach to compute an optimal environment model for the desired application based on the available sensor data. In particular, this includes the following main topics:

- Sensor setup
- Sensor data fusion
- Self-diagnosis
- Driver intent estimation

Initially, a suitable high performance sensor setup is to be defined. Based on the sensor setup, a reliable environmental perception approach is to be derived, which should be able to perform a continuous diagnosis of the provided data even with respect to unforeseeable events (e.g. degraded view due to bad weather conditions, damages on sensor hardware, sensor decalibration or aging) to allow a dynamical functionality adaptation and predictive system limits recognition. Additionally, it should be possible to estimate the intents of the detected traffic participants within this approach.

A maximum synergy between the individual driver assistance functions should be capitalized and therefore an integrated approach with central management of sensor data should be chosen. This allows the systems to make maximal use of the available data and additionally assures the entire system consistency due to the common information core.

1.4 Overview of the thesis

Chapter 2 recapitulates the current state of the art concerning the research field addressed by this thesis. Thereby series products as well as the most recent research topics are referred. It results in a problem definition, which reflects the objectives of this thesis and sets it apart from the related work. A compact theoretical background, that is needed for better understanding of subsequent chapters, is summarized in chapter 3. It focuses on general methods designed for handling of uncertain knowledge, discusses their particular suitability for the given objectives and gives reasons for the choice of Bayesian networks. Moreover, various types of Bayesian networks are presented and the fundamental mathematical background is illustrated by means of an example.

Based on previous chapters, a general probabilistic approach to multiple sensor environmental perception of advanced driver assistance systems is derived in chapter 4. Chapter 4 contains the main scientific contributions of this thesis: In particular, section 4.2 describes a generic sensor data fusion mechanism, its probabilistic representation by a Bayesian network and a novel application of the derived principle by means of a probabilistic track-to-track data fusion developed especially for modern sensors with built-in internal tracking. The new high-level self-diagnosis capability, which assigns a probabilistic confidence level to the fused objects, is explained in detail in section 4.3. It reflects the main domains of uncertainty resulting from the object's state variables, from its general detection as well as from its association. The principle of maneuver level intent estimation of detected objects using an example of a lane change maneuver is presented in section 4.4. The objective of the section 4.5 is to present the integration of the probabilistic methods introduced in section 4.2, section 4.3 and section 4.4 to form a robust environmental perception mechanism with self-diagnosis and intent estimation ability. The unique full integration of the particular methods is possible due to the uniform modeling technique choice. This improves significantly the resulting system consistency and allows further extensibility.

In chapter 5, the experimental system is introduced which was prototypically developed to demonstrate and approve the concepts described in this thesis, whereby the section 5.2 focuses on its hardware and software architecture. In chapter 6 individual components of the developed system are evaluated by means of real test scenarios, and the according results are presented and discussed. The results of the new approach are very promising due to the sensor data fusion results precision, and the relatively high percentage of well assigned situations by the self-diagnosis. Finally, chapter 7 presents the summary of this thesis and an outlook.

Chapter 2

State of the art

The aim of road traffic safety systems is to reduce the harm (number of fatalities, injury severity level, or property damage) resulting from collisions of road vehicles. Generally, one can distinguish between active and passive safety.

Passive safety systems provide safety if an accident is just occurring and their task is to minimize its consequences. We speak about active or preventive safety in case of foresighted accident avoidance or mitigation systems. Such systems try to avoid or mitigate potential collisions before they happen, based on sensorial information about the vehicle's environment [Bau02].

While the passive safety measures (e.g. energy absorption materials, seat belts, airbags) have reached a high maturity level in the recent decades, the active safety domain still offers an unexploited potential according to the today's experts' opinion. Furthermore, the passive safety systems can contribute to the reduction of accident consequences, but they cannot prevent

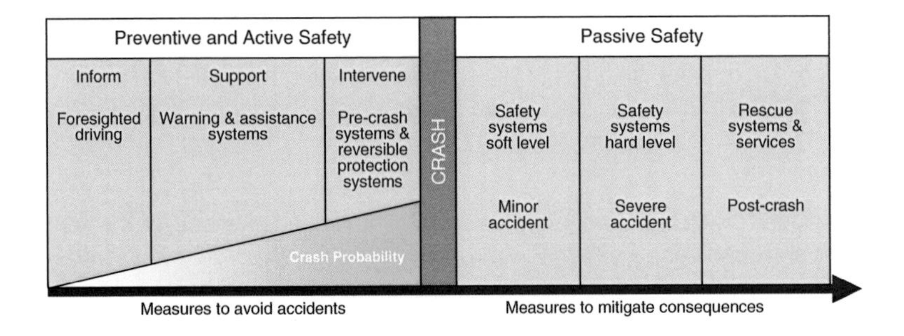

Figure 2.1: Active and passive safety measures

accidents. Thus the active safety, in adequate combination with passive safety, promises a further reduction of hazard during road traffic.

Figure 2.1 gives an overview about the road traffic safety measures in relationship to the remaining "time to collision" (TTC) [SMI+08].

Driver assistance systems (DAS) support the driver in his tasks. Basic driver assistance systems consider the host vehicle's internal sensors (e.g. odometry) for this purpose, as for example the anti-lock braking system (ABS), the electronic stabilization program (ESP) or the brake assist system (BAS) do. These systems support the driver in the control of his vehicle especially in hazard situations. Nowadays they are already available as a standard equipment of all vehicle classes down to mini class segment.

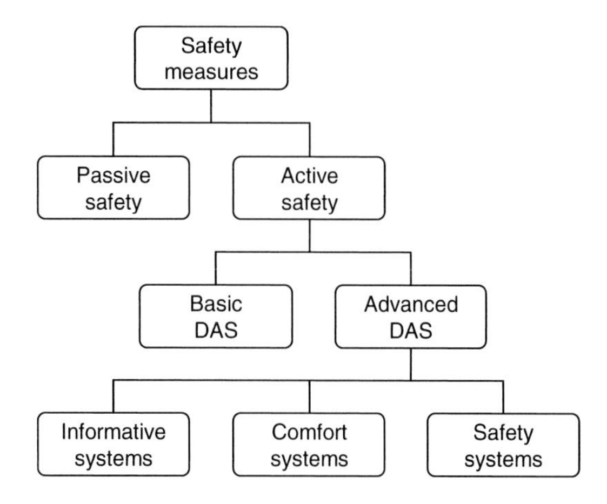

Figure 2.2: Hierarchical overview of safety measures

Figure 2.2 depicts the internal structure of the domain of safety measures. Besides the basic driver assistance systems, an important subset of active safety measures is represented by the advanced driver assistance systems.

Advanced driver assistance systems (ADAS) additionally observe the vehicle's surroundings by external environmental sensors (based e.g. on radar, laser or video camera principle). Besides the state of the host vehicle the information about other traffic participants, about the driving lanes or about additional traffic infrastructure has to be captured and processed. This information is used in order to inform the driver (e.g. about potential hazards) or even directly intervene in the vehicle dynamics. Hence, hazard situations can be recognized in their very early stage and potential collisions can be avoided or at least significantly mitigated.

2.1 Advanced driver assistance systems

In the active safety domain, the advanced driver assistance systems represent a significant safety potential [SMI+08, GEJS08]. Depending on the support level, the ADAS can be subdivided into [Stü04]:

- Informative systems which inform of hazard situations and warn of potential driving faults.
- Comfort systems which relieve the driver and make the driving-task more convenient.
- Safety systems which avoid collisions or at least reduce the consequences.

Informative systems recognize potentially dangerous situations and possible faults of the driver of the host vehicle. Such systems inform the driver in advance. However, they do not directly influence the vehicle dynamics. An example is the lane departure system, which gives the driver an advice of accidentally leaving the driving lane.

Comfort systems support the driver through his routine tasks. An example is the adaptive cruise control (ACC). In the most current ACC series systems radar sensors are used to measure the distance, the relative speed and the azimuth of the target vehicle. Initiated by the driver, the distance regulation sets continuously the speed of the host vehicle according to the sensed information, so that a secure distance (time gap) to the target vehicle is ensured.

Safety systems are supposed to avoid accidents or – if the accident is no more avoidable – to reduce its consequences. They intervene automatically using the actors of the host vehicle. An example of such system is an automatic collision-avoiding maneuver or automatic brake intervention [Kop00]. Thereby a significant challenge is to recognize the accident is definitely unavoidable by the driver, especially if irreversible actions are to be done.

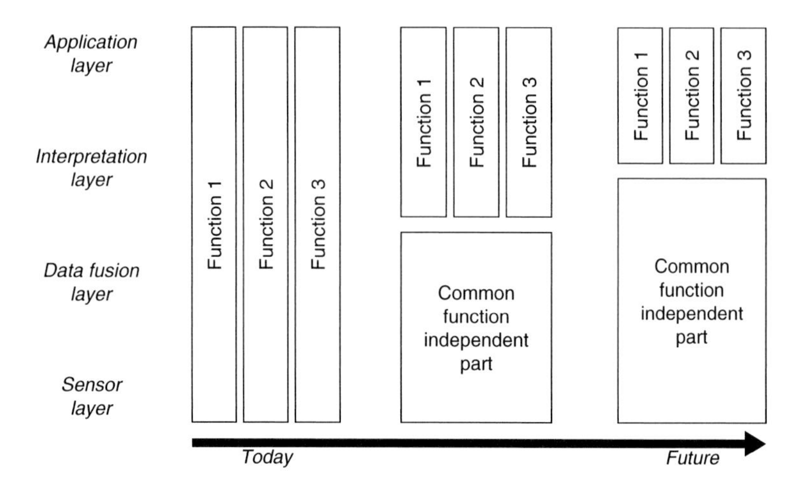

Figure 2.3: ADAS environment perception development with increasing common function independent part in the future

Driver assistance systems of the different levels have different requirements and aims. Safety systems require for example higher technical availability than comfort systems.

The number of ADAS is growing continuously, especially in the last few years. ADAS are no more a domain of research projects. In the meantime, a variety of ADAS is well-established on the market as well. According to the research area addressed by this thesis (s. section 1.3) this section focuses mainly on ADAS designed for the use on highways, which support the driver in lateral and/or longitudinal direction.

Although the individual ADAS differ significantly, their topology can be generalized using the following components:

- Environment perception (sensor and data processing)
- Controller application (of actors and HMI)
- Human-machine interface (HMI)
- Actor(s)

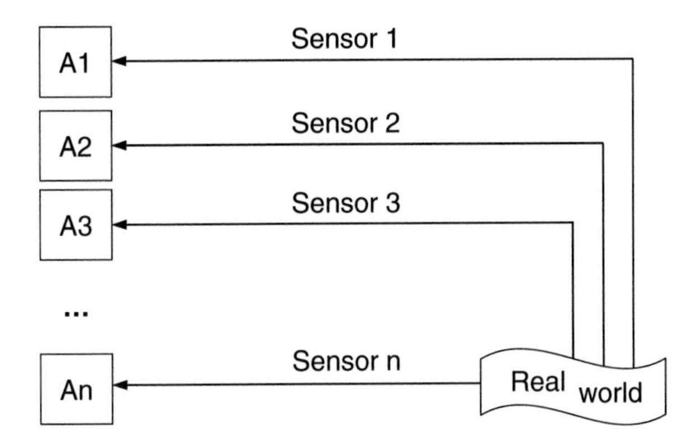

Figure 2.4: ADAS architecture overview: separated perception; An – individual driver assistance applications; Sensor n – according sensors

2.1.1 Series products

The most of today's advanced driver assistance systems (e.g. adaptive cruise control) are based on an exclusive sensor or sensor set (s. figure 2.4). They are generally separated, so the potential synergy effects remain widely unused. In most cases the entire processing chain from the sensor up to the HMI triggering is integrated in a single control unit.

According to the opinion of today's experts, the common function independent part of advanced driver assistance systems will be further increasing. Figure 2.3 depicts this phenomenon.

ADAS supporting the driver (directly or indirectly) in lateral and/or longitudinal direction, which are available on today's market, are introduced in this section.

Advanced Cruise Control

Advanced Cruise Control (ACC) represents the first ADAS introduced on the European market[1]. The ACC automatic distance control builds on the

[1]Originally introduced in Europe by Daimler (Distronic) in 1998: 77GHz radar based system, field of view 3x 3 degrees, range about 80m.

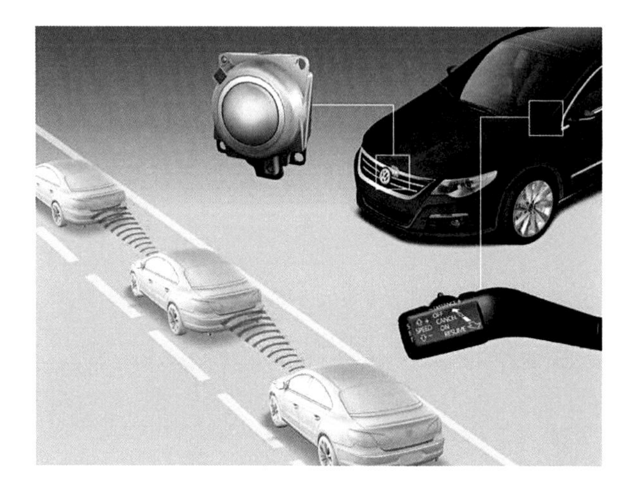

Figure 2.5: Advanced cruise control: system overview (source: VW)

well-known cruise control system by adding an environment sensor (e.g. radar). When activated, the system maintains a preset time gap from the vehicle ahead by automatically adjusting the speed of the host vehicle accordingly (s. figure 2.5). The environment sensor and control module are usually combined into a single unit.

Using the signals from the radar sensor, the control module computes not just the distance to the vehicle ahead but also the vehicle's relative speed, as well as its lateral offset on multi-lane carriageways. If there are several vehicles within the sensor's field of coverage at the same time, this information is used to select which of the vehicles the system follows.

If approaching a slower vehicle ahead or another vehicle cuts in front, the system slows down the car by initiating corrective controls in the engine management and, if necessary, in the braking system as well. If the required rate of deceleration exceeds a defined maximal amount of the vehicle's braking power, the driver is prompted to apply the brakes manually.

The original ACC system is being further developed and extended (e.g. "follow to stop", "stop and go" functionality)

Figure 2.6: Lane Assist: system overview (source: VW)

Lane Departure Warning and Heading Control

The lane keeping assistance systems can help avoid the accidents caused by vehicles that deviate from their lane. Such systems inform the driver about the potential hazard[2] (e.g. by vibration of steering wheel or driver seat) or even directly perform a correcting action by intervening in the steering system of the host vehicle to support the driver before the vehicle starts to leave its lane[3] (s. figure 2.6).

When the system is activated, on-board road sensor detects the lane markings and evaluates the pose of the vehicle relative to the lane course. If the car tends to drift off lane, the lane keeping assistance system informs the driver or performs the correcting action. The system does not react if the vehicle moves out of its lane after the driver has indicated. The active steering technology is available in vehicles equipped with electronically "steerable" electro-mechanical power steering (EPS) without use of additional actor.

[2]Originally introduced in Europe by Citroën (Lane Departure Warning) in 2005: based on infrared light sensors mounted under the front bumper, feedback by vibration of the appropriate side of driver's seat.

[3]Originally introduced in Europe by Volkswagen (Lane Assist) in early 2008: based on CMOS camera behind the windshield, the system automatically countersteers as soon as the host vehicle starts to leave its lane unintentionally.

2.1.2 Research projects

Research projects result from the cooperation of automotive industry and public institutions. Public institutions regularly support automotive industry cooperations in activities which are of common public interest. Hence, co-funded research project with the aim of road safety can be supported.

On the one hand, the research projects from the past represent a historical preamble of today's series systems. On the other hand, the state of the art of today's research projects represents an outlook of today's series systems.

By means of selected projects dealing with topics addressed by this thesis the current state of this research area will be described. Both European international and German national projects are taken into account.

CARSENSE (2000–2002)

Within the CARSENSE project (Sensing of Car Environment at Low Speed Driving) an ACC system was extended for complex situations and low speed range. A monocular camera system was used to estimate the lane course, while the data of a stereo camera system and laser scanner were used for the object recognition. A basic sensor data fusion mechanism was used on the one hand for the allocation of objects relative to a lane, and, on the other hand, to reduce potential errors due to occlusion of lane markings caused by environment objects. CARSENSE started in 2000 and the project duration was 2 years [Lan01, CFBM02].

INVENT (2001–2005)

The objective of INVENT (Intelligent traffic and user-oriented technology) was to optimize traffic flow and enhance traffic safety. The project was divided in more research areas. One of them was called "driver assistance and active safety". It dealt with wide spectrum of applications from lateral assistance and traffic jam assistance on highways to intersection assistance for cities. The experience from this project showed that individual sensor systems are not sufficient for handling the information required in complex traffic situations. For this reason, a further focus of research work was the networking of various sensor systems (radars, laser scanners, video cameras) to form a driving environment sensing system. Hence, an approach to generate a comprehensive electronic image of the driving environment in the test vehicle itself was introduced, allowing a deep online analysis of the traffic situation. This four-year cooperation project took place through mid 2005 [Wei07].

PReVENT (2004–2008)

PReVENT (PReVENTive and Active Safety Applications) was started to develop, test and evaluate a wide spectrum of safety-related applications, using advanced sensors, positioning and communications devices integrated into on-board systems for driver assistance. One of the research areas focused on the lateral support of a vehicle to help the driver to keep his vehicle at the safest position in the lane, as well as warn the driver if he is about to run off the road. Furthermore some research work was dedicated to the field of sensors and sensor data fusion techniques used for ADAS. This four year project started in 2004 [SMI+08].

AKTIV (2006–2010)

AKTIV is the abbreviation used for "Adaptive and cooperative technologies for intelligent traffic". Its objective was the development of solutions for enhanced traffic safety and optimal efficiency of traffic flow. This research initiative was subdivided into more main projects dealing with traffic management, car-to-car communication and active safety. Within the subproject "Integrated Lateral Assistance" an ADAS is being developed, designed to continuously support the driver in steering and lane-keeping on highways, well-built country roads, and through narrow construction zones in both lateral and longitudinal direction. Due to the assisting functionality, the driver was always allowed to intervene and override the current system support if required.

Precise knowledge of the vehicle environment constituted an essential prerequisite for the integrated lateral assistant application to act foresightedly in a wide spectrum of traffic situations. Therefore various sensors were used to gather data about the area around the vehicle, and the perception system fused the information from all of the individual sensors into a consistent, comprehensive model. This environmental information was

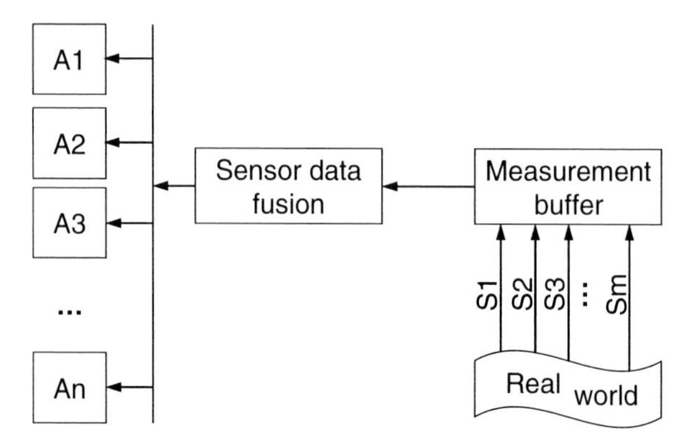

Figure 2.7: Common perception; An – individual driver assistance applications; Sm – individual sensors

processed by the assistance application where an appropriate control signal for the actors was generated [GEJS08]. The particular challenges of this subproject have been addressed by this thesis (s. section 1.1).

2.2 Sensor data fusion

As stated in section 2.1, future advanced driver assistance systems need comprehensive information about the environment of the host vehicle to fulfill their tasks. Single individual sensor systems and sensor technologies are no longer able to deliver the entire required information (with the desired availability); rather a network of different sensors becomes necessary. The Figure 2.7 shows a bundle of assistance systems with a common environmental perception (cp. figure 2.4 – separated perception).

In the case of a common perception instead of transferring the sensor data directly to the assistance application the data is fused and interpreted by an additional sensor data fusion module. This module contains a virtual environmental model – an information source for the entire assistance applications.

Common environment perception allows the assistance applications to use the information about vehicle's environment of all sensors. Due to the more extensive environment information, which can be reached by a system with a common environment perception, the existing assistance systems can be extended in their functionality and/or availability. Furthermore, by sharing of resources the complexity and the costs of the overall system can be reduced compared to several individual systems. And finally, the compatibility of the actions of different assistance applications must be ensured. A common environment perception ensures consistent information base for all used assistance applications.

Depending on the sensor setup, one of the following sensor data fusion architecture types can be generally implemented [Stü04]:

Redundant fusion The data-acquisition is realized by several identical sensors with the same field of view. The accuracy of the observation can be increased through supplementary values (reduction of uncertainty). Furthermore an increased robustness against individual sensor failures is reached (a malfunction of a sensor restricts neither the common field of view nor the feature room of the observations).

Complementary fusion One speaks about complementary fusion if the field of view of the fused sensors does not contain any cross-over regions of more sensors or if the observed features distinguish. Through the combination of the sensor data either the total field of view or the feature space of the observation is extended. In this case, a failure of individual sensors always results in loss of information.

Cooperative fusion Cooperative data fusion combines observations from different sensors to produce new features (e.g. triangulation of two distance-measuring sensors). Only by the combination of more measurements the derived feature can be determined.

In practice, the fusion architecture is often a mixture of more of the described architecture types, e.g. the fields of view and features are partly identical and partly complementary.

The resulting environmental model can vary as well. The methods described in the literature can be divided generally into two different branches:

Object based approach Classically the real environment of the host vehicle is greatly simplified and reduced to a relatively small number of relevant objects (e.g. traffic participants) whose dynamics is then individually estimated [Kal60, JR07].

Grid based approach Another possibility is to decompose the environment in a defined amount of cells, where each cell represents the occupancy of the associated environment area. Although this idea is known for more decades [ME84], it is a subject of today's intensive research [NMM+08].

Sometimes a combination of both environment model representations can bring advantageous results [NMM+09, Eff09]. In the following the basic principles of both methods are described.

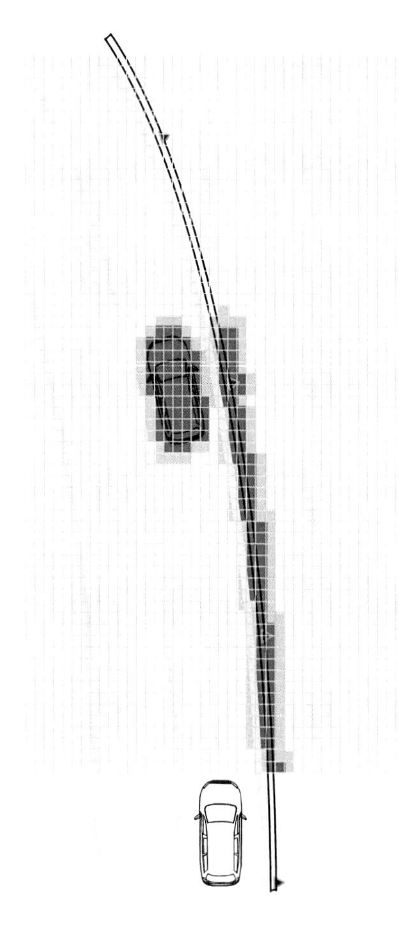

Figure 2.8: Grid based approach

2.2.1 Grid based approach

Unlike the object-based techniques, the grid-based techniques use a fine-grained grid card models to describe the host vehicle's environment [ME84, MM96, NMM+08]. Every grid cell carries information about the occupancy status of the corresponding environmental space, which allows an explicit modeling of open areas between obstacles. Each of the cells represents a small segment of the real environment in the resulting occupancy grid. Every single measurement influences the occupancy probability in the according grid area. Thus, there is a direct connection between the space captured from a measurement and the update of the information in the grid. Various sensor measurement types can be incorporated directly into the grid. Higher logical processing (e.g. object recognition) is not necessary at this level. The level of detail is determined by the resolution of the grid and the information content of individual cells (feature space). The basic feature space would be binary, which can only distinct between free and occupied cells. The memory space grows over a square with the size of the modeled environmental area. Furthermore, the complexity increases over a square with the grid resolution. Objects are not explicitly modeled, therefore information bound to objects (e.g. object type) cannot be reproduced at this level. Furthermore only spatial and not spatiotemporal qualities can be represented by a grid. A prediction of the future environment setup is therefore not possible. This kind of representation suits well especially for unstructured stationary environments (e.g. urban).

2.2.2 Object based approach

In case of object-based environment modeling the host vehicle environment is described by a set of detected objects. Each object is defined by a set of properties (e.g. position, velocity, size). These properties are generally centralized in a so called state vector. For each object an object model is given. The contents of the state vector depend on the model specification. Based on the state vector and additional model assumptions about the movement of the objects, their "future" can be predicted – more exact their future movement and hence their state vector after a certain time Δt. This is reconstructed through the use of classical state estimation techniques in the context of the so-called object tracking from the available data. By increasing the prediction step length the quality of the prediction degrades, because the used mathematical model cannot reproduce all aspects of the real object's movement. The environment model bases on the list of the detected objects, which is the basis for all high-level interpretation tasks. The dimension of the objects state vectors determines the complexity of the environment model, and hence the degree of simplification. As an example, the point-target representation of the vehicles can be sufficient for a number of applications. The complexity of the modeling increases linearly with the total number of detected environmental objects. This kind of modeling is suited especially for well structured environments (i.e. highway).

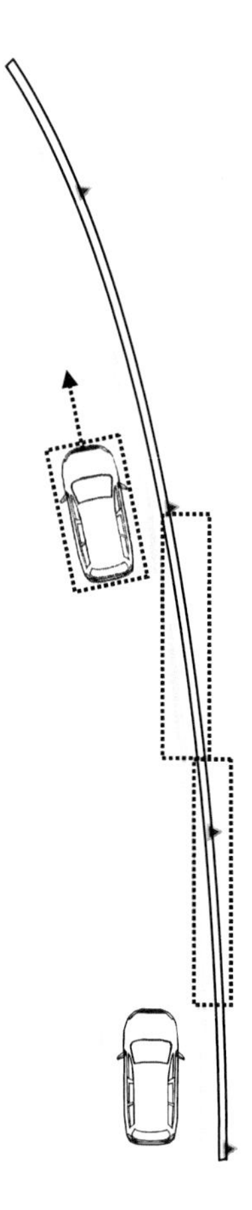

Figure 2.9: Object based approach

With respect to the given constraints (s. section 1.3) this approach will be generally followed in this thesis, whereby more alternatives will be considered, as described in following.

Depending on the sensor output data an appropriate object based fusion algorithm variant can be chosen. The classical measurement data fusion is designated for direct (noisy) measurements. A so-called track-to-track fusion can be advantageous in case of tracked data, which is the common output of today's automotive sensors.

Although the perspectives of individual algorithms differ, conceptually all of them try to obtain an optimal estimate of desired quantities from data provided by a noisy environment [May79].

Measurement data fusion

The measurement (or sensor-to-track) fusion of more heterogeneous sensors is achieved generally by an recursive asynchronous event based filtering algorithm, which updates its state every time a new measurement arrives. The sensors send their measurements to the central unit which processes them. Thereby the solution with the minimal deviation among a defined class of estimators is selected. The main drawback of this method is the need for raw measurements and the necessary low level implementation, which is tightly coupled with the hardware resources (camera, radar, etc.).

A state vector of a tracked object $\mathbf{x}_k$ at time k can be calculated as follows:

$$\mathbf{x}_k = f(\mathbf{x}_{k-1}, \mathbf{u}_{k-1}) + \mathbf{q}_{k-1} \tag{2.1}$$

Thereby $\mathbf{x}_{k-1}$ represents the state vector at time $k-1$ and $\mathbf{q}_{k-1}$ the process noise vector at time $k-1$. Control vector $\mathbf{u}_{k-1}$ represents e.g. the driver actions in the tracked target vehicle of interest. It is usually unknown and therefore neglected.

The above equation is related to a so called prediction step: $f(\mathbf{x})$ represents the state transition model (which is applied to the previous state). The following equation corresponds to the update step after a sensor measurement $\mathbf{z}_k$ at time k has arrived:

$$\mathbf{z}_k = h(\mathbf{x}_k) + \mathbf{r}_k \tag{2.2}$$

Thereby, $\mathbf{r}_k$ describes the observation noise at time k and $h(\mathbf{x})$ the observation (sensor) model which maps the true state space into the observed space.

The above equations represent the mathematical basis for many filtering algorithms (e.g. Gaussian filters, nonparametric filters) as described in [WB95, May79, RN03] and practically evaluated e.g. by [CFBM02, Wei07, SMI+08].

Gaussian filters This class of filtering algorithms is leaded by the well-known Kalman filter [Kal60]. Although the original approach is limited to the consideration of linear systems with normally distributed stochastic noise processes, it has been successfully used for various applications since several decades. Its main advantage is the existence of a closed algebraic solution. Thus, $f(\mathbf{x})$ and $h(\mathbf{x})$ from equation 2.1 and equation 2.2 are linear functions and the functions result in a matrix multiplication with $\mathbf{F}$ resp. $\mathbf{H}$. Furthermore $\mathbf{q}_k$ and $\mathbf{r}_k$ are normally distributed with the according covariance matrix $\mathbf{Q}$ resp. $\mathbf{R}$. With respect to the described relations, the state vector estimate $\hat{\mathbf{x}}_k$ can be expressed as follows:

$$\hat{\mathbf{x}}_k = \hat{\mathbf{x}}_k^- + \mathbf{\Sigma}_k^- \mathbf{H}^T (\mathbf{H}\mathbf{\Sigma}_k^- \mathbf{H}^T + \mathbf{R})^{-1} \cdot (\mathbf{z}_k - \mathbf{H}\hat{\mathbf{x}}_k^-) \tag{2.3}$$

where following terms are used:

$$\hat{\mathbf{x}}_k^- = \mathbf{F}\hat{\mathbf{x}}_{k-1} \tag{2.4}$$

$$\mathbf{\Sigma}_k^- = \mathbf{F}\mathbf{\Sigma}_{k-1}\mathbf{F}^T + \mathbf{Q} \tag{2.5}$$

The first term of equation 2.3 represents the predicted state, the middle term describes the optimal Kalman gain and includes the innovation covariance and the last term represents the innovation whereby $\mathbf{H}_k\hat{\mathbf{x}}_k^-$ refers to the predicted measurement $\hat{\mathbf{z}}_k^-$. The according covariance matrix $\mathbf{\Sigma}_k$ is defined as follows:

$$\mathbf{\Sigma}_k = \mathbf{\Sigma}_{k-1} - \mathbf{\Sigma}_k^- \mathbf{H}^T (\mathbf{H}\mathbf{\Sigma}_k^- \mathbf{H}^T + \mathbf{R})^{-1} \mathbf{H}\mathbf{\Sigma}_{k-1} \tag{2.6}$$

In practice, a linear process model is often not met. However, inserting of a normally distributed variable in a nonlinear function does not results in a normally distributed variable and thus the filtering algorithm would terminate.

The so-called "extended Kalman filter" (EKF) addresses this issue using the following linearized Jacobian matrices at time k:

$$\mathbf{F}_k = \left.\frac{\partial f(\mathbf{x})}{\partial \mathbf{x}}\right|_{\mathbf{x}=\hat{\mathbf{x}}_k} \text{ and } \mathbf{H}_k = \left.\frac{\partial h(\mathbf{x})}{\partial \mathbf{x}}\right|_{\mathbf{x}=\hat{\mathbf{x}}_k} \tag{2.7}$$

By this linearization of the observation and state transition models, the problem is reduced to the original Kalman filter (see above) and the remaining algorithm works analog.

The Jacobian matrices $\mathbf{F}_k$ and $\mathbf{H}_k$ are to be calculated at each time step, which implies a correspondingly higher computational complexity.

Due to the linearization the EKF provides only approximations to the state vector and its covariance. The size of the error depends on the nonlinearity degree of the process model. There are two main sources of this error [DVDD97]:

Truncation Error This error is based on the linearization of the fact that only the first derivative is considered and the higher order terms cut off. The addition of these terms is possible, but leads to an increased computational effort.

Base Point Error The linearization is always done at the estimated value, not the true value. This leads to the linearization conditional estimation error, this can further negatively influence the filter convergence.

In order to avoid the effects of the EKF linearization a so called "unscented transformation" can be alternatively used. The background of this method is the idea that it is easier to approximate a normal distribution than an arbitrary non-linear function or transformation [JU97]. The resulting "unscented Kalman filter" (UKF) represents another way to deal with the linearization problem. Instead of approximating the function $f(\mathbf{x})$ by partial derivatives (as in equation 2.7), the algorithm selects specific points, which represent the covariance matrix, and passes them through the original function $f(\mathbf{x})$ to reconstruct the expected value and the transformed covariance matrix. That means no linearization has to be computed. For n-dimensional random variable an amount of $2n+1$ so called sigma-points is to be selected. The details of this algorithm can be found in [JU97].

Nonparametric filters Nonparametric filters are generally able to deal with arbitrary distributions and/or non-linear functions. The resulting distribution is approximated by a finite number of values – samples. Increased number of samples increases the approximation quality, but simultaneously the computational complexity as well. One important member of this class is the Particle Filter [HM69, GSS93, Thr02]. This algorithm can be described as follows [TBF05]:

- Initially, all particles are individually predicted by means of an arbitrarily configurable process model according to the equation 2.1.
- Each particle is weighted according to its deviation to the currently received measurement vector. Any arbitrary noise process can be considered within the context of the according sensor model.
- Then the particle cloud is rebuilt. Thereby the particles are duplicated according to their weight in the rearrangement. Particles with low weight can be removed. The resulting cloud of particles approximates then the desired distribution of system state.

In order to reduce the computational complexity of this class of algorithms they are sometimes combined with Gaussian methods for selected components. This kind of approach is known as "Rao-Blackwellisation", because it is related to the Rao-Blackwell formula originally presented by [AK77].

Track-to-track data fusion

The measurement fusion bases on a central node which processes the entire sensor measurements (s. previous section) is doubtless the theoretically optimal method. Nevertheless, todays automotive sensors usually output already processed object lists based on single sensor data. The details of sensor preprocessing and tracking are usually kept confidential by automotive suppliers. Even if a raw data interface can be provided, hardware-specific proprietary sensor knowledge is necessary to gain full advantage of such sensor data. Sensor fusion based on already pre-tracked sensor objects is therefore an eligible type of automotive data fusion [SMI+08].

Sometimes measurement data fusion algorithms (s. previous section) are "misused" to fulfill this task. This approach is called pseudo-measurement data fusion and it neglects the temporal correlation of the sensor data. Despite its theoretical suboptimality, this approach can lead to satisfying results in practice [WSK03]. Nevertheless, applying a Kalman filter to filtered data leads coercively to additional delays and to general underestimation of the fused objects covariance due to temporal correlations of individual sensor data [MA08].

The global estimate $\hat{\mathbf{x}}$ and its covariance $\mathbf{\Sigma}$ of two sensors A and B with the local and external trackers estimates, $\hat{\mathbf{x}}_A$ and $\hat{\mathbf{x}}_B$, and their estimation error covariance matrices $\mathbf{\Sigma}_A$ and $\mathbf{\Sigma}_B$ can be expressed as functions of the following variables:

$$\hat{\mathbf{x}} = f(\hat{\mathbf{x}}_A, \hat{\mathbf{x}}_B, \mathbf{\Sigma}_A, \mathbf{\Sigma}_B, \mathbf{\Sigma}_{AB}) \tag{2.8}$$

$$\mathbf{\Sigma} = f(\mathbf{\Sigma}_A, \mathbf{\Sigma}_B, \mathbf{\Sigma}_{AB}) \tag{2.9}$$

Correlation suppression This method respects the temporal correlation of individual sensor data and neglects the common process noise between the tracks of more sensors. The global estimate according to this method is a convex combination of the local estimates of the individual sensor data without a need of further inputs [SK71]. In the following the equation for two sensors A and B with the according trackers estimates $\hat{\mathbf{x}}_A$ and $\hat{\mathbf{x}}_B$, and their covariance matrices $\mathbf{\Sigma}_A$ and $\mathbf{\Sigma}_B$ is presented:

$$\hat{\mathbf{x}} = \mathbf{\Sigma}_B(\mathbf{\Sigma}_A + \mathbf{\Sigma}_B)^{-1}\hat{\mathbf{x}}_A + \mathbf{\Sigma}_A(\mathbf{\Sigma}_A + \mathbf{\Sigma}_B)^{-1}\hat{\mathbf{x}}_B \tag{2.10}$$

$$\mathbf{\Sigma} = \mathbf{\Sigma}_A(\mathbf{\Sigma}_A + \mathbf{\Sigma}_B)^{-1}\mathbf{\Sigma}_B \tag{2.11}$$

Even though this method leads to good tracking accuracy and prevents additional delays, it is mathematically not completely correct, because the tracks are actually correlated through their common process noise.

Weighted covariance This method was developed in order to take the correlation between the local tracks in account [BS81]. In the following the equation for two sensors A and B with the according trackers estimates $\hat{\mathbf{x}}_A$ and $\hat{\mathbf{x}}_B$, and their covariance matrices $\mathbf{\Sigma}_A$ and $\mathbf{\Sigma}_B$ is presented:

$$\hat{\mathbf{x}} = \hat{\mathbf{x}}_A + (\mathbf{\Sigma}_A - \mathbf{\Sigma}_{AB})(\mathbf{\Sigma}_A + \mathbf{\Sigma}_B - \mathbf{\Sigma}_{AB} - \mathbf{\Sigma}_{AB}^T)^{-1}(\hat{\mathbf{x}}_A - \hat{\mathbf{x}}_B) \tag{2.12}$$

$$\mathbf{\Sigma} = \mathbf{\Sigma}_A - (\mathbf{\Sigma}_A - \mathbf{\Sigma}_{AB})(\mathbf{\Sigma}_A + \mathbf{\Sigma}_B - \mathbf{\Sigma}_{AB} - \mathbf{\Sigma}_{AB}^T)^{-1}(\mathbf{\Sigma}_A - \mathbf{\Sigma}_{AB})^T \tag{2.13}$$

where $\mathbf{\Sigma}_{AB}$ represents the so called cross-covariance matrix, a measure of the tracks correlation due to common process noise. It is computed recursively. However, for its computation the details of both sensors Kalman filters (such as state-transition and dynamic models) have to be known. This is usually not the case, especially for industrial sensors with integrated tracking. For such case, an additional estimation algorithm of the cross-covariance matrix is needed [BSL95].

Covariance intersection Covariance intersection method was proposed to combine tracks of unknown correlation [HL01]. It deals with the problem of invalid incorporation of redundant information and it produces a comparatively conservative solution. In the following the equation for two sensors A and B with the according trackers estimates $\hat{\mathbf{x}}_A$ and $\hat{\mathbf{x}}_B$, and their covariance matrices $\mathbf{\Sigma}_A$ and $\mathbf{\Sigma}_B$ is presented:

$$\hat{\mathbf{x}} = \mathbf{\Sigma}(\omega\mathbf{\Sigma}_A^{-1}\hat{\mathbf{x}}_A + (1-\omega)\mathbf{\Sigma}_B^{-1}\hat{\mathbf{x}}_B) \tag{2.14}$$

$$\mathbf{\Sigma}^{-1} = \omega\mathbf{\Sigma}_A^{-1} + (1-\omega)\mathbf{\Sigma}_B^{-1} \tag{2.15}$$

where $\omega \in [0,1]$ has to be determined through an optimization process to minimize a defined criteria of uncertainty (e.g. $\det(\mathbf{\Sigma})$ [JU07]). However, this method can fail in case of spurious estimates, which have to be concerned separately [SMI+08].

Covariance union The covariance union approach allows to unify two tracks, even if the difference of the state estimates exceeds the covariance indicated by at one of the tracks [Uhl03]. This method guarantees consistency as long as the system and measurement estimates are each consistent. In the following the equation for two sensors A and B with the according trackers estimates $\hat{\mathbf{x}}_A$ and $\hat{\mathbf{x}}_B$, and their covariance matrices $\mathbf{\Sigma}_A$ and $\mathbf{\Sigma}_B$ is presented:

$$\hat{\mathbf{x}} = \arg\min(\det(\mathbf{\Sigma})) \tag{2.16}$$

$$\mathbf{\Sigma} = \max(\mathbf{\Sigma}_A + (\hat{\mathbf{x}} - \hat{\mathbf{x}}_A)(\hat{\mathbf{x}} - \hat{\mathbf{x}}_A)^T, \mathbf{\Sigma}_B + (\hat{\mathbf{x}} - \hat{\mathbf{x}}_B)(\hat{\mathbf{x}} - \hat{\mathbf{x}}_B)^T) \tag{2.17}$$

As for the covariance intersection method an optimization process has to be performed in order to solve the above equations.

2.3 Self-diagnosis

In context of today's vehicles the originally medical term "diagnosis" is used for accurate mapping of findings about failures in the on-board electrical network and its sub-systems.

The first section of this chapter presents an overview of diagnostic methods used in the automotive domain, with emphasis on advanced driver assistance systems and the according environmental perception. The second section focuses especially on uncertainty and existence estimation of objects detected by environmental sensors.

2.3.1 Automotive diagnosis

The on-board electrical system of a modern vehicle contains a number of subsystems, controllers and other components with increasing trend of complexity. A search for an error cause in such systems would be practically impossible without accessing the internals of the system. This is, on the one hand, the reason for introducing "diagnosable" car controllers[4]. On the other hand, diagnosable emission control systems has been legislatively forced since late 80's.[5] These requirements were conditional on a common standard, which is today referred to as On-Board Diagnostics (OBD) [EC98].

The on-board diagnostic functionality of today's vehicles covers far more than the monitoring of the emission relevant systems. It allows access to further manufacturer specific data, which are provided by the built-in software of each control device (s. e.g. section 2.1.1). Eventual errors are stored in the respective controller as so called diagnostic trouble codes (DTC) and, depending on its severity, a control indicator (e.g. lamp) might be switched on to inform the driver. The error entry in the controller is a basis for further off-board failure analysis in the corresponding workshop.

On- and Off-board diagnosis

Depending on its application, one can distinguish between on-board and off-board diagnosis methods.

Certain information must be available immediately after the occurrence of an error in the vehicle. Problems in the system are recognized independently and further communicated (e.g. control lamp). For this purpose so-called on-board diagnosis systems are used.

On the other hand, information about the cause of an error (detected by the on-board diagnosis) necessary for its elimination are relevant only for the workshop. For this purpose complex off-board diagnosis systems can be used.

A close cooperation between on-board and off-board diagnosis is needed to achieve optimal results in general. This thesis deals exclusively with on-board diagnosis methods.

[4]Originally introduced by Volkswagen (Type 3) in 1969.

[5]Started in California for all new passenger vehicles in 1988.

Diagnosis in the context of environmental perception of ADAS

In the domain of advanced driver assistance systems the common environmental perception methods can provide an information about uncertainty of an object in addition to its state estimate (e.g. variance, s. section 2.2). This can be treated as a kind of confidence measure of the according environmental object(s) – a basic continuous diagnostic measure of the system at a different level than conventional on-board diagnostics (s. section 2.3.1).

Therefore, complementary to the conventional diagnostics, an online internal analysis of the examined system is proposed, which estimates the confidence level of the output data based on a comprehensive analysis of the input data. This can give a hint, if the system is able to solve the desired task under given conditions – a continuous feedback of system functionality, where low level indicates the system limits are currently likely reached and does not necessarily mean a system defect.

This idea can be compared with a so-called "quality of service" (QoS), used mainly in the field of telephony or networking. Quality of Service comprises requirements on all available aspects of a connection, such as service response time, loss, signal-to-noise ratio, cross-talk, echo, interrupts, frequency response, loudness levels, and others.

High and low level diagnosis

Nearly all modern controllers monitor their own operations. If an error is identified, an according diagnostic trouble code is stored in the designated error memory. The entries in the error memory can simplify the search for the possible cause of the problem. This well-established kind of diagnosis is in this thesis referred to as "low-level" diagnosis. It is principally based on proof of the plausibility of the signals of sensors and other components (e.g. broken wire). It is designed for failure detection with limited dynamics, and rudimentary reliability monitoring (e.g. static or sporadic).

On the other hand, the so-called "high-level" diagnosis introduced in this thesis focuses on cross-system continuous analysis, which reflects a confidence level of defined application outputs (e.g. lane markings). Depending on its result, the following application can (temporarily) reduce the available functionality. In extreme case, it can cause a defined function to be (temporarily) completely disabled. Instead of an error entry in memory, the system functionality can be immediately adapted. Table 2.1 gives a brief comparison of both levels of diagnosis.

	Initial step	Typical action steps
Low level diagnosis	input signals plausible	no action
	input signals implausible	generate DTC, enter emergency operation mode, request maintenance
High level diagnosis	analyze input data, long term monitoring	output the estimated confidence level of the provided data to client application(s) $\rightarrow$ initiate application specific functional adaptation (e.g. give advice instead of intervene)

Table 2.1: Comparison high-level and low-level diagnosis

High- and low-level-diagnosis are to be considered as complementary. Successful low-level diagnosis is a prerequisite of high-level diagnosis. Satisfactory results can be reached only by combination of both levels. In the last years, probabilistic methods for low-level-diagnosis get popular [KBL$^+$07]. This allows further coupling of both described levels. This thesis focuses on high-level diagnosis.

2.3.2 Uncertainties and existence estimation

Environmental perception of modern advanced driver assistance systems bases on sensor data fusion methods (s. section 2.2). There are the following general uncertainty domains in context of sensor data fusion [DKK05]:

- State variables uncertainty
- Detection uncertainty
- Association uncertainty

The most runtime confidence estimation algorithms consider exclusively the uncertainty of the state variables. The following sections are structured in accordance to the addressed sources of uncertainty.

State variables uncertainty

Most of the approaches presented in section 2.2 make use of the uncertainties resulting additionally to the objects' state estimations directly from the used filtering algorithm. They can be treated directly as a basic quality measure of the environmental perception. This principle can be applied for detected environmental objects and for the host vehicle as well [Wei07]. Hence in case of Gaussian algorithms, the error covariance matrix $\mathbf{\Sigma}$ of the estimated object state vector $\mathbf{x}$ can be used as a rudimentary measure of the estimated state confidence (s. equation 2.18). For n state vector components the covariance matrix is $n \times n$ sized:

$$\mathbf{x} = \begin{bmatrix} x_1 \\ x_2 \\ \vdots \\ x_n \end{bmatrix}; \; \mathbf{\Sigma} = \begin{bmatrix} var(x_1) & cov(x_1,x_2) & \cdots & cov(x_1,x_n) \\ cov(x_2,x_1) & var(x_2) & \cdots & cov(x_2,x_n) \\ \vdots & \vdots & \ddots & \vdots \\ cov(x_n,x_1) & cov(x_n,x_2) & \cdots & var(x_n) \end{bmatrix} \tag{2.18}$$

The covariance $cov(x_i,x_j)$ is a measure of correlation of the according state vector components. Special case represents $x_i = x_j$, where

$$cov(x_i,x_j) = cov(x_i,x_i) = var(x_i) \tag{2.19}$$

Variance $var(x_i)$ is a measure of "spread" of a variable x_i from its expected value or mean μ_i. Main diagonal of covariance matrix is often a subject of the main interest. It contains variances of x_i. If $x_1...x_n$ are completely independent, all matrix elements except the main diagonal would be zeros. Variance differs in units from the original variable and therefore it makes sense to introduce its square root as the standard deviation σ_i, where

$$var(x_i) = \sigma_i^2 \tag{2.20}$$

Statistically, the interval $\mu_i \pm \sigma_i$ contains 68% values of a normally distributed random variable x_i and the value of σ_i can be therefore used as a confidence indicator.

One property of variance resulting from a filtering algorithm is its independence of μ_i; it depends solely on the individual variances of the source data and not on the according means. In practice, the expressiveness of σ_i can be therefore limited.

Detection uncertainty

The detection uncertainty can be expressed by means of existence probability of individual objects tracked by an arbitrary sensor data fusion algorithm. Track existence probability P_{ex} calculation at time k is based on the following general recursive formula:

$$P_{ex}(k) = f[P_{ex}(k-1), Q] \tag{2.21}$$

where the sensor specific probability ratio Q can influence P_{ex} positively, negatively or neutrally depending on current sensor data, as explained below.

In [Bla86], a well-established existence estimator is introduced, known as Bayes estimator. The primary purpose of this algorithm was a multiple hypothesis tracking, assuming conditional independence of consecutive measurements. This method is based on the following recursive formula:

$$P_{ex}(k) = \frac{P_{ex}(k-1)}{P_{ex}(k-1) + Q \cdot [1 - P_{ex}(k-1)]}; \quad Q \in \{Q^-, Q^0, Q^+\} \tag{2.22}$$

Depending on the sensor field of view and the measurement association success of delivered sensor data three possible values of ratio Q are supposable as illustrated in table 2.2.

	Data association	Field of view	Effect on P_{ex}
Q^-	not successful	inside	decreasing
Q^0	not successful	outside	no change
Q^+	successful	—	increasing

Table 2.2: Possible values of ratio Q

An extensive method is presented in [MRD06] to jointly address the tracking and the existence probability issue. The method is demonstrated by means of two sensors – laser and camera. The projected laser echoes generate region of interest for an image sub-window classifier. The algorithm utilizes the detection box cluster size (number of boxes it contains) and the laser echo amplitude feature to estimate the objects existence probability.

Association uncertainty

The so-called association distance represents a statistical measure of association uncertainty commonly used to assign sensor measurements to individual objects in the pre-stage of a tracking algorithm.

Association distance can be used as quality measure of dynamic model suitability as well [Wei07]. In this case an unsuitable dynamic model is assumed as the a priori cause of its increased value. According to this approach the best-suited dynamic model or a weighted combination of more models can be selected for tracking purposes.

The basic measure could be e.g. the geometrical Euclidean distance e_k of both vectors at time k:

$$e_k^2 = \sum_{i=1}^{n} (\mathbf{z}_{k_i} - \hat{\mathbf{z}}_{k_i}^{-})^2 \tag{2.23}$$

where $\mathbf{z}_k$ is the sensor measurement and $\mathbf{z}_k^{-}$ the predicted theoretical measurement based on a dynamic model. This approach does not incorporate the according uncertainties and is therefore suboptimal. The so-called Mahalanobis distance d_k solves this issue [Mah36] for normally distributed components. Its square, the so-called normalized innovation squared (NIS), can be calculated as follows:

$$d_k^2 = (\mathbf{z}_k - \hat{\mathbf{z}}_k^{-})^T \cdot \mathbf{S}_k^{-1} \cdot (\mathbf{z}_k - \hat{\mathbf{z}}_k^{-}) \tag{2.24}$$

This equation contains the term $\mathbf{z}_k - \hat{\mathbf{z}}_k^{-}$, which expresses the difference between the predicted and the measured data (innovation), and additionally the according covariance matrix $\mathbf{S}_k$.

2.3.3 Further methods

Within the research project PReVENT (s. section 2.1.2) a theoretical framework was investigated with the objective to get a reliable confidence information about the environmental model at runtime. They defined confidence as a function that, given a model of some object inferred from feature data or aggregated from other objects, provides a value, which reflects the representation quality of the model with respect to the observed data. Thus, one confidence value expressed the entire model quality. Furthermore a fast evaluation technique for the derived confidence measure was given using the so called inference trees was presented, a graphical method to draw a logical tree by applying classical inference rules (e.g. Modus Ponens). [SMI+08]

2.4 Intent estimation of traffic participants

The environmental perception of modern driver assistance systems provides input data for assistance applications. This data bases on information from different sources fused to a common environment model as described in section 2.2 in detail. The assistance applications usually request a defined subset of this model, possibly specified at high abstract level (e.g. ACC relevant object). Thus the environmental perception must be able to allow an additional interpretation of the produced model. This forms the base for the intent estimation of individual traffic participants. Thereby, different levels of intents can be considered (cp. section 2.1): The enumeration starts from elementary intents and ends with more complex (abstract) intents.

- Handling-Level
- Maneuver-Level
- Navigation-Level

The handling-level covers all elementary actions of the driver which deal with longitudinal and lateral vehicle control (e.g. steering, braking and accelerating). This is in accordance with the application field of basic DAS (section 2.1). The maneuver-level includes the interaction of the driver with other traffic participants according to the domain of ADAS (e.g. lane change). The navigation-level deals with high abstract driving tasks (e.g. follow the desired route, reach destination).

In general, there is often an anxious distinction between the estimation of the intents of the host vehicle and intents of other detected traffic participants, whose state is being estimated within the environmental model. Although, there is less precise information of the state of the other vehicles available than about the state of the host vehicle, a universal approach is followed in this thesis. The only difference between the host vehicle (ego) and other vehicles is the accuracy, and eventually the partial (un)availability of individual state variables, e.g. the yaw rate and eventually the yaw angles of the environmental vehicles are usually not measured directly, but estimated from the movement of the external vehicles under given assumptions.

This section focuses on methods designed for intent estimation of the host vehicle or other traffic participants in well-structured environments (e.g. on highways) according to the objectives of this thesis.

2.4.1 Multiple-model filtering

Basically, it is possible to extend a given sensor fusion approach by additional dynamic models. Thus, multiple prediction hypotheses can be derived for a given object.

Within such multi-model filter algorithm, arbitrary maneuver hypotheses can be represented by partial models. Hence, it is possible to directly compare the hypotheses by their model likelihoods. The model likelihoods can be estimated directly within the filtering framework. Based on the association distance values (s. section 2.3.2) and the set of dynamic models describing the maneuvers of interest, the best-suited dynamic model for the current maneuver of the individual environmental object can be chosen. Thereby the computation of further measures is not necessary. The primary purpose of this type of approach is the handling-level intent estimation of environmental objects. Based on this information, conclusion about maneuver level intents can be made.

[Wei07] makes use of a multi-model technique, which incorporates and continuously evaluates two maneuver hypotheses of lane keeping and lane change, in order to recognize a lane change maneuver of other traffic participants. Therefore both hypotheses were implemented as part models of an multiple-model filter. In fact, the proposed model set contains two models of constant velocity for longitudinal and lateral movement. The lane change intent is derived from the relative lateral dynamics of the observed environment object. Similar strategy was implemented by [AS06]. They incorporate further attributes (e.g. direction indicator) and the estimation result is based on continuous actual-theoretical values comparison.

The detection of potential lane change intent of the host vehicle can be implemented by use of different association hypotheses in the sensor model of a lane sensor. Based on the comparison of predicted and measured features, an certain tendency can be concluded [Wei07].

2.4.2 Time to lane change

Based on a given vehicle state, an according dynamic model and a driving lane description, the time to the future moment when a vehicle is going to cross one of the adjacent lane markings of the driving lane can be analytically determined. This time gap is commonly called time to lane change (TLC).

Prediction of the lane change based on the determination of TLC requires very precise movement and lane course estimation. Thus it is primary used for the host vehicle [KLK98, RME00].

Under assumption of velocity persistence of the host vehicle (constant velocity model) and a clothoidal lane description with constant curvature (s. section 4.2.3), the t_{LC} can be calculated according to [Wei07] as follows:

$$d_{LC_{1-4}} = \frac{-tan\psi \pm \sqrt{tan^2\psi - 2(\frac{\dot{\psi}}{v} - \kappa)(y_0 \pm \frac{b}{2})}}{\frac{\dot{\psi}}{v} - \kappa} \tag{2.25}$$

$$t_{LC} = d_{LC_{min}} \frac{\sqrt{1 + tan^2\psi}}{v} \tag{2.26}$$

where the included symbols can be interpreted as follows:

$d_{LC_{1-4}}$	[m]	distances to the possible lane crossing hypotheses
ψ	[rad]	yaw angle of the vehicle relative to the lane axis
$\dot{\psi}$	[rad/s]	yaw rate of the vehicle
v	[m/s]	absolute velocity of the vehicle
κ	[1/m]	curvature of the lane
y_0	[m]	initial vehicle offset according to the lane axis
b	[m]	lane width

2.4.3 Further methods

A fuzzy system designed for overtaking maneuver prediction of the host vehicle was proposed by [BSF08]. It based on raw data available on the vehicle internal bus, including object data from an ACC sensor and extended navigation data. The main inputs were: brake pressure, accelerator pedal position and velocity, distance to the next intersection, distance to ACC target and its relative velocity. The estimated intents were: turn, overtake or follow.

Similar approach was presented by [SMKY07]. He used a complex psychological driver model based on a theory for simulating and understanding human cognition. He made use of following inputs: steering angle, accelerator pedal position, lateral position in the lane, front vehicle time gap, rear approaching vehicle(s) time gap, information about neighbor lanes.

A situation assessment for the detection of cutting-in vehicles in order to enhance present ACC-Systems is presented in [Dag05]. His concept bases on a common ACC system. It considers uncertainty arising from the inaccuracy of sensor data by means of a probabilistic network. He uses a pure diagnostic approach (s. section 3.3.1), where the observed effects imply the hidden variable, which represents the cutting-in vehicle.

The intent estimation of the host vehicle from driver's point of view has been examined in detail by [SG08]. A general method based on probabilistic networks is proposed that allows systematical evaluation of the situation's characteristics and estimate the driver's future actions. The resulting probabilistic network has a causal topology. The influence factors are divided into two groups: motivators, which can cause the driver to do a defined maneuver, and indicators, which signal the defined maneuver as a consequence. The approach is exemplarily demonstrated by means of overtaking intent recognition.

The research project PReVENT (s. section 2.1.2) addressed the topic of intent estimation as well [SMI+08]. They analyzed both the behavior of environmental objects and of the host vehicle and made a classification according to a predefined discrete set of classes (e.g. overtaking, lane change, lane drifting, etc.). They assumed the existence of an interpreted environment model (containing descriptions of the road attributes and the lane properties and the output of the objects' path and the lane assignment). They tried to find the most appropriate motion of a vehicle in a given environment by solving an optimal control problem. For this purpose, the optimal plan called "reference maneuver" was proposed to suggest a safe way of driving. Its computation was based on a goal function expression that had to be minimized. The optimal plan, being the "best" in the given conditions and scenario, was used as a reference to derive the intents of a driver.

2.5 Problem definition

Table 2.3 summarizes selected approaches presented in this chapter together with their kind of coverage of the essential topics of this thesis. Thereby general approaches addressing broader research field are preferred to specialized approaches addressing individual topics. The columns reflect the particular objectives of this thesis specified in section 1.3. The individual rows of table 2.3 present the name of the corresponding system, project or author and brief description of the used data fusion, diagnosis and intent estimation methods. Additionally, the basic application of each approach is included. The integrability of individually treated interrelated topics is generally desired. Nevertheless, due to e.g. the variety of themes and/or the project size the global approach consistency and therefore the possible integration level can be limited. This aspect is expressed by the last column of table 2.3.

A combination of adaptive cruise control (ACC) and lane assist (LA) represents a quite sophisticated ADAS set available on today's market. It bases on separated perception architecture: each function obtains its data exclusively from the own sensor. Thus, the vehicles can be equipped with both systems independently of each other due to no common parts. Nevertheless, the potential synergy effects of both systems remain unused. Within the INVENT research project [Wei07], a common perception approach was implemented, which included a kind of diagnosis measure by means of the state variables uncertainty. Furthermore, an analytical intent estimation method based on time to lane change (TLC) prediction and a set of dynamic models was realized. The main limit of this approach comes from the neglect of detection and association uncertainty as a part of the diagnosis measure and the limited extensibility. Within the framework of PReVENT project [SMI+08], numerous algorithms had been developed including an extensive common perception, a hierarchical confidence level estimator based on inference trees and an intent estimator based on a defined reference maneuver. Due to the global approach comprehensiveness and heterogeneity its consistency and the possible integration is limited. Maehlisch & Dietmayer developed a measurement data fusion algorithm which jointly addresses the state and the existence estimation problem [MRD06]. This algorithm supports a full speed ACC application. It focuses especially on state variables and detection uncertainty. Nevertheless, the combination of the uncertainty sources to one common measure goes beyond their scope as well as an intent estimation algorithm.

The last row of table 2.3 reviews the objectives of this thesis as specified in section 1.3. The presented method bases on a common general instrument to model the addressed system components. Thus, a consistent and simultaneously an extensible approach can be implemented. It incorporates an object based sensor data fusion mechanism, online confidence level estimation with respect to all sources of uncertainty identified in section 2.3.2 and additionally a possibility to interpret the results by means of intent estimation of detected traffic participants. Moreover, it allows a natural incorporation of an arbitrary a priori information.

Approach	Data fusion	Diagnosis	Intent estimation	Application	Integration level
ACC and LA	separated perception	low level (OBD)	—	basic lateral and longitudinal support	independent systems
INVENT	common perception; measurement fusion	state vector variances	dynamic model fitting; TLC	wide spectrum of ADAS	partial integration
PReVENT	common perception; object and grid based fusion	environment model confidence level using inference trees	reference maneuver	wide spectrum of ADAS	independent algorithms
Maehlisch & Dietmayer	measurement fusion	existence probabilities	—	full speed ACC	full integration
This thesis	measurement and track level fusion (chapter 4.2)	high level self-diagnosis (chapter 4.3)	probabilistic inference (chapter 4.4)	advanced lateral and longitudinal assistance (chapter 5)	full integration (chapter 4.5)

Table 2.3: Matrix overview of the problem definition

Chapter 3

Theoretical Background

In the domain of environmental perception of driver assistance systems a number of high sophisticated mathematical models and algorithms is used as presented in chapter 2. In case of a complete knowledge about the environment, a pure application of logical methods would be guaranteed to produce optimal results. Unfortunately, in reality this is not the case[1]. Any model of real phenomenon is always incomplete due to the presence of some hidden variables. Thus, one has to face the problem of incompleteness, which implies uncertainty [BLS08]. In section 2.3.2 the problem of uncertainty was already introduced. In this chapter, general methods are presented, which are specially designed to act under uncertainty. This research area is often referred to as soft computing. Its main objective is to exploit the tolerance for imprecision, uncertainty, partial truth, and approximation to achieve tractability, robustness, low solution cost and better rapport with reality [Zad94].

3.1 Soft computing methods

Soft computing forms a contrast to the so-called hard computing, the exact data processing, based on clearly defined data and precise calculation rules and conclusions. Opposite to this, soft computing works with blurred knowledge, incomplete data and inaccuracies. It is not one homogeneous approach, rather a partnership of distinct methods, which can be roughly divided in approximate reasoning, and functional and optimization approximation areas, including search methods as depicted in figure 3.1. This section provides a brief overview over the methods of soft computing.

[1] Albert Einstein's quote on this topic: "As far as the laws of mathematics refer to reality, they are not certain, as far as they are certain, they do not refer to reality."

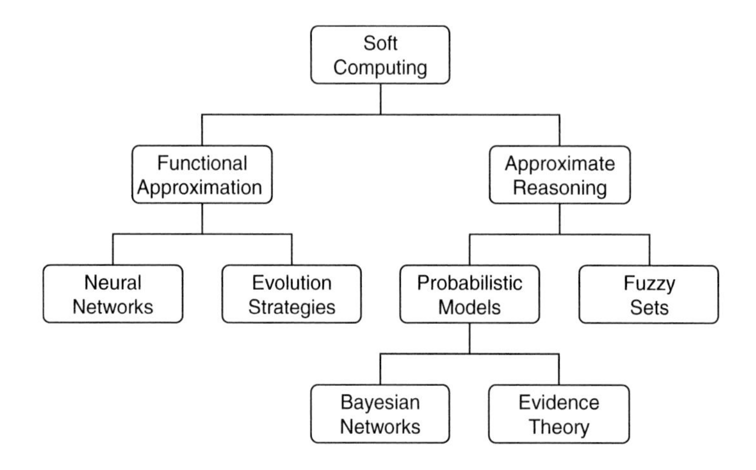

Figure 3.1: Overview: methods of soft computing

3.1.1 Functional approximation

Functional approximation bases on the use of any other functional representation then a table. It is referred to as "approximate" because the original function cannot be represented in a closed form in the most cases. Figure 3.1 shows the components of functional approximation area – neural networks and evolutionary algorithms – introduced in the following.

Neural networks

Neural networks [Ros57] try to "imitate" the structure of nerve cells with the aim of modeling, learning or classifying. This is usually performed based on known examples from which a generalized model is calculated. Neural networks consist of units, based on a basic mathematical model of a neuron [MP43]. The units are connected by directed arcs – links, that are used to propagate eventual activation from one unit to another. Furthermore, links have numeric weights assigned, that determine the strength and the sign of the connections. The output a_x of a single neural unit x depends on the linear combination of its inputs a_i and the according weights $W_{i,x}$.

Assuming a unit with n inputs, its output can be derived as follows:

$$a_x = f\left(\sum_{i=0}^{n} W_{i,x} a_i\right) \tag{3.1}$$

where $f(x)$ represents the activation function of the unit, based usually on a defined hard or soft threshold.

Evolution strategies

The evolution strategies [Rec73] were originally introduced to solve optimization problems in the aerospace domain (airfoils). They are principally based on targeted random search, inspired by biological mechanisms (e.g. adaptation, evolution). Therefore they can help to understand the background of those mechanisms as well. Their strength is especially in optimization and adjustment tasks. Thereby, solely an evaluation function has to be set up to allow rating of single solution candidates. Based on this function a search or optimization can be performed. Among the evolutionary algorithms are e.g. genetic algorithms. Genetic algorithms make use of so-called stochastic beam search. The general algorithm can be described as follows:

- It starts with n random states (population).
- Each state i is evaluated with a so-called fitness function $f(i)$. The "better" the state, the higher value this function returns.
- Then, a subset of the states is selected in accordance to their rankings, with the probability of choosing a given state i being an increasing function $p(f(i))$ of its ranking.
- The members of the chosen subset are paired and the successors are built by crossovering of parents definitions with a random crossover point.
- Finally, the children definitions can be additionally randomly muted with a low occurrence probability.

Genetic algorithms combine random exploration with the exchange of information among threads. Their power comes mainly from the crossover operation, where even bigger independently evolved blocks can be combined. The way they work causes genetic algorithms sometimes to be seen more as living organisms than solutions.

3.1.2 Approximate reasoning

Besides the functional approximation techniques, approximate reasoning represents another set of powerful methods to deal with uncertainty (s. figure 3.1).

Fuzzy set theory

Fuzzy set theory [Zad65] is based on a vague concept, which is used as a basis for approximate reasoning. The essential idea is to switch from the classical two-valued notion of truth to gradual, multi-valued notion of validity of statements or decisions. The gradual truth value is mathematically described by the theory of fuzzy sets, which provide a possibility to define a degree of satisfying a hypotheses. Fuzzy set is a set without sharp boundaries, where a predicate is implicitly defined by a set of its members.

Fuzzy logic deals with logical terms describing membership in fuzzy sets. Conclusions regarding a specific problem can be derived using modus ponens as inference method. Fuzzy logic is based on following standard rules for the evaluation of fuzzy truth T:

$$T(A \wedge B) = \min(T(A), T(B)) \tag{3.2}$$

$$T(A \vee B) = \max(T(A), T(B)) \tag{3.3}$$

$$T(\neg A) = 1 - T(A) \tag{3.4}$$

Fuzzy techniques are well-suitable for specifying, how well an object satisfies a defined vague description, which does not necessarily result from uncertainty about the external world. Despite the (theoretical) complete knowledge about the object, its assignment to the defined description can still be unclear.

Probabilistic models

This section examines the modeling of uncertain knowledge using probabilistic methods such as Bayesian networks and the evidence theory (s. figure 3.1).

Bayesian networks BN [Pea88] provide a compact graphical representation of probabilistic models. They allow explicit modeling of causal dependencies. A Bayesian network is represented by a directed acyclic graph, where the nodes describe system variables and the directed edges possible conditional dependencies between them. Each node possesses a conditional distribution, which bases on its parent nodes. The uncertainty of a node is represented by a probability distribution for a state assumption and its alternatives.

Once a BN is specified, it is able to compute an arbitrary set of query variables based on an optional set of evidence variables. This operation is called (probabilistic) inference, an update mechanism where the probabilities are propagated causally in the network.

Mathematical background of probabilistic inference is based on the generalized form of the Bayesian rule:

$$P(S|K) = \sum_{\forall F} \frac{P(S,K,F)}{P(K)} \tag{3.5}$$

The resulting conditional probability $P(S|K)$ is represented by summation over all free variables, where S stands for "Searched", K for "Known" and F for not specified "Free" variables. An example inference calculation can be found e.g. in section 3.4.

Bayesian networks provide an intuitive visual representation of conditional (in)dependence relations of the system. Furthermore, they allow an efficient representation of the full joint distribution, which includes all combinational values of all system variables (cp. section 3.2).

Evidence theory The evidence theory [Dem67, Sha76] (or Dempster-Shafer Theory) is a probabilistic method for combining hypotheses of variable credibility. Evidence theory can handle uncertainty and ignorance simultaneously, and distinguish between them. Instead of computing directly a probability of a hypothesis, it calculates the probability that the evidence supports the hypothesis. This measure of belief is denoted as $Bel(H)$ [RN03]. In case of no evidence for all defined hypotheses H_i, the according belief value can be expressed as follows:

$$Bel(H_i) = 0 \tag{3.6}$$

That means no belief in any of the hypotheses, independently of their theoretical probability of occurrence! This gap arises due to full ignorance. If some evidence e_0 related to the hypothesis H_x appears, which causes the general system confidence of $C(e_0)$:

$$Bel(H_x) = C(e_0)P(H_x) \tag{3.7}$$

Alternatively, a belief interval can be defined as follows:

$$Bel(H_x) \leq P(H_x) \leq 1 - Bel(\neg H_x) \tag{3.8}$$

The width of the interval expresses the measure of ignorance, which can help to decide if more evidence is needed.

3.1.3 Discussion

Functional approximation methods are generally able to represent functions for large state spaces. Their significant benefit is the ability to generalize from known data to further unknown data. On the other hand, these methods fail, if there is not any function in the chosen hypothesis space to approximate the true original function sufficiently well. Models based on functional approximation are difficult to ensure a stable deterministic behavior. In this regard such models behave black-box-like. Despite their potential suitability, they rarely become accepted in the automotive domain due to this behavior. These methods are predestined for offline processing, because of their ability to inductively generalize over the input states and relatively slow convergence [RN03].

Fuzzy sets theory is a way to represent vague information. The strength of this concept is the ability to perform operations based on this information which can still lead to reliable results. In the domain of driver assistance systems, the use of fuzzy logic has been investigated in relation to vehicle control algorithms in detail [TMS+08]. Because vagueness and uncertainty are in fact orthogonal problems, truth-functional methods are generally not well suited for representing uncertain reasoning. Vagueness does not necessarily result from uncertainty about the environment. Its complete description can be available, but an assignment to a given property can still be unclear.

The evidence theory is mainly suitable for tasks where some uncertain (with ignorance afflicted) information possibly from different sources should be combined. A significant part of the applications described in literature deal with classification problems. In the environmental perception domain, the evidence theory is used in cases, where ignorance and non-evidence must be separately considered. This is the case especially for grid based techniques (s. section 2.2.1). Unfortunately, the evidence theory does not allow a definite decision in many cases, where other probabilistic methods yield a specific choice [RN03].

Bayesian networks provide a well-developed representation of uncertain knowledge with causal topology, intuitive architecture and fast extensibility. They accommodate a well balanced trade-off between expressiveness and tractability. Based on Bayesian networks, a wide variety of well-known (probabilistic) models can be implemented that have proved to be successful in practice as demonstrated in the section 3.2. The restriction to Bayesian networks still provides quite a lot of freedom in modeling. Even the meaning of ignorance can be expressed by means of change of a belief in accordance to a new evidence (cp. equation 3.8).

The techniques introduced in this section are based on different principles and have all their strengths and they focus on various areas. Therefore, they should be seen as complementary rather than competitively.

With respect to the objectives of this thesis (s. section 1.3), Bayesian networks were chosen as a general instrument to model the addressed system components. Thus, a consistent approach can be implemented incorporating an object based sensor data fusion mechanism, an online confidence level estimation of its result and additionally a possibility to interpret the results by means of intent estimation of detected traffic participants. This integrated probabilistic approach is derived in chapter 4.

3.2 Bayesian networks

Figure 3.2 shows the hierarchy of well-known probabilistic models [Mur02, DBM03]. It starts from more general models in the top and continues to more specific models in the bottom. This fact implies that the more specific models can be expressed by the more general models (e.g. a continuous Hidden Markov Model (HMM) represents a special case of a dynamic BN).

Bayesian networks (BN), originally introduced by Judea Pearl in [Pea88], provide a compact representation of the probability distribution of all involved variables taking advantage of known conditional independences and additional a priori knowledge. Bayesian networks are directed acyclic

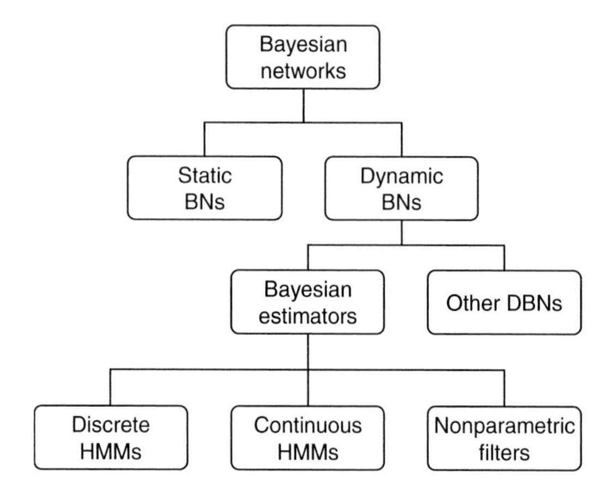

Figure 3.2: Bayesian networks: hierarchy of selected probabilistic models

graphs whose nodes represent variables, and whose edges represent conditional dependencies among them (figure 3.3). Typically, BNs are depicted in causal direction only. The direction of inference (dotted arrow) is shown here for once to illustrate this relation.

Nodes can represent either discrete variables with finite number of mutually exclusive values (section 3.2.1) or continuous variables with defined distributions (section 3.2.2) or a combination of both section 3.2.3. A conditional probability table or a density function (for continuous nodes) is assigned to each variable (node). A priori knowledge is assigned to parentless nodes. The nodes can be observable and in this case evidences can be considered and the network can be used to find out the state of the unobserved variables. This process of computing the posterior distribution of variables given evidence is called inference. Additionally to equation 3.5, it makes use of a so-called chain rule, which applied on a Bayesian network consisting of a set of nodes X_i can be calculated as follows:

$$P(X_1,...,X_n) = \prod_{i=1}^{n} P(X_i|parents(X_i)) \tag{3.9}$$

Furthermore, this rule represents a base for probabilistic networks evolving over time (for time series modeling), called dynamic Bayesian networks (DBN).The use of dynamic Bayesian networks allows incorporation

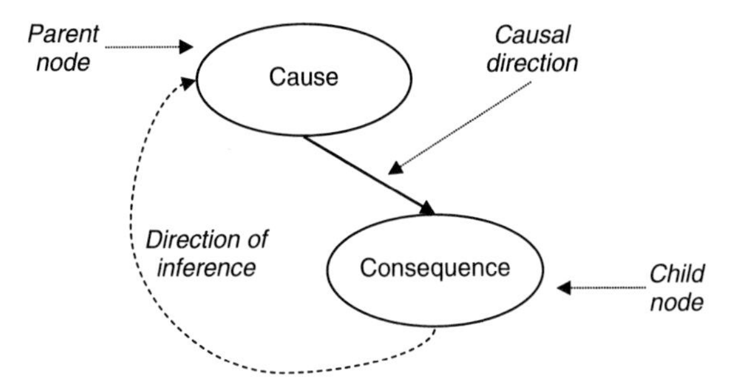

Figure 3.3: Bayesian networks: basic principle

of "memory" into the models. In this way a conditional dependence of a variable on e.g. its last state can be defined. Thereby discrete time-development is assumed, increasing the time index t every time a new data (observation) is available. The network remains in accordance to its original definition. Hence, its topology does not change over time (s. section 3.2.4).

3.2.1 Discrete Bayesian networks

Discrete BNs consist of a set of discrete or discretized variables. Each random variable possesses a set X of possible discrete values x. Its initial value is specified by the prior probability $P(X=x)$ of x, which describes the according degree of belief, if no other evidence has been observed.

Once another information (evidence) about another variable $Y=y$ is available, prior probabilities are no more valid. Instead, one has to respect the so-called conditional probability $P(X=x)$ conditioned on $Y=y$, written as $P(X=x|Y=y)$. It can be interpreted as the probability that variable X takes on value x, given that all one knows is that variable Y has been observed to be y. This is also called posterior probability and abbreviated as $P(x|y)$. The relationships between prior and conditional probabilities can be expressed as:

$$P(X|Y) = \frac{P(X,Y)}{P(Y)} \quad \text{and}$$
$$P(X,Y) = P(X|Y)P(Y) \tag{3.10}$$

where $P(Y) > 0$. By a substitution of X through Y in the above equation and combining both equations together, the original Bayesian theorem can be derived:

$$P(X|Y) = \frac{P(Y|X)P(X)}{P(Y)} \tag{3.11}$$

By substituting Y through $A \cap B$ and reducing the resulting fraction a so-called conditional Bayesian theorem can be derived:

$$P(X|A,B) = \frac{P(A|X,B)P(X|B)}{P(A|B)} \tag{3.12}$$

Conditional probabilities in discrete BN are specified for each possible value of each node explicitly in local conditional probability table of that node (CPT). Such table specifies the dependencies between the parent nodes and itself. The size of this table is proportional to the number of parents and to the number of possible values of a node. Joint probability distribution describes a CPT for a combination of a set of more random variables.

3.2.2 Continuous Bayesian networks

In practice, many systems contain continuous quantities (e.g. distance, velocity). Continuous variables have theoretically infinite number of possible values. It is not possible to define conditional probabilities for each value. Thus the conditional distribution cannot be represented by a CPT.

One possibility in such case is to divide continuous variables into a discrete set of intervals – a so-called discretization. A variable does not necessarily have to be discretized linearly: the area of interest can be e.g. described in more detail then the rest. The experiences have generally shown it is possible to reach good results in spite of coarse granularity.

Another possibility is to define the probability that a random variable X takes on some value x as a parametrized function of x, which can be specified by a finite number of parameters. Such probability distributions are called probability density functions (PDF). Gaussian distribution, defined by its mean μ_x and its variance σ_{xx}^2, is a popular PDF. There is a slight discrepancy between the meaning of $P(X = x) = a$ for both the discrete and the continuous case:

Discrete case Probability, that X takes on the value of x. The result is a dimensionless number a.

Continuous case According to the definition for discrete case $P(X = x)$ would be theoretically always zero, because the continuous variable X will never take on the exact value x. It rather means that the probability of X being in the small region dx around x is in the limit equal to a divided by the region width in the appropriate units.

$$P(X = x) = \lim_{\Delta x \to 0} \left. \frac{P(x \leq X \leq x + \Delta x)}{\Delta x} \right|_x = \frac{a}{unit} \tag{3.13}$$

The result of this equation is no more dimensionless, it is measured by reciprocal units (if existent).

3.2.3 Hybrid Bayesian networks

Systems modeled by hybrid BNs provide a possibility to incorporate both discrete and continuous variables in a common model. Table 3.1 summarizes all possible combinations of a parent and child nodes. Additionally

parent node	child node
discrete	discrete
discrete	continuous
continuous	discrete
continuous	continuous

Table 3.1: Possible parent-child node combinations

to previous definitions, it is necessary to specify the process to define the following distributions:

Continuous variable with discrete parents Discrete parents of continuous nodes can be handled by explicit enumeration as in the case of discrete child nodes. One basically needs to specify the child node distribution for all possible values of the parent.

Discrete variable with continuous parents In this case, one needs to specify how the values of the discrete variable depend on the distribution of the parent. This can be handled by a threshold function.

A possible adequate soft threshold function is the so-called probit distribution [RN03], which bases on the integral of the standard normal distribution.

3.2.4 Dynamic Bayesian networks

Dynamic Bayesian networks are generally based on static Bayesian networks. The additional time component allows one to pass information from the past to the future states. Z^t represents a set of entire variables Z_i of the system at time t. The system specification consists of two nets. The first one is static, defining the prior $P(Z^1)$ and the second one is temporal with the time transition $P(Z^t|Z^{t-1})$ between two instances of the static network defined as follows:

$$P(Z^t|Z^{t-1}) = \prod_{i=1}^{N} P(Z_i^t|parents(Z_i^t)) \tag{3.14}$$

Each node in the second instance of the static network has an associated conditional probability distribution (CPD) of $P(Z_i^t|parents(Z_i^t))$ for $t > 1$. The network structure is in accordance with the Markov assumption. Under this assumption, the resulting joint distribution is defined as follows [Mur02]:

$$P(Z^{1:T}) = \prod_{t=1}^{T}\prod_{i=1}^{N} P(Z_i^t|parents(Z_i^t)) \tag{3.15}$$

3.3 Architecture of Bayesian networks

A fully specified BN provides a complete description of the system. It can be used to answer any question based on the full joint distribution of involved variables. Network specification consists on the one hand of the general topology and on the other hand of the conditional probabilities for each node.

System topology specifies the conditional dependence and independence of variables inside the system and if applicable the degree of dependence. This can help to understand the system and it is usually more definite than specifying conditional probabilities, which describe the details of the dependences including their strengths.

3.3.1 Topology

Based on equation 3.9 the conditional dependencies of a network are partially restricted. In combination with equation 3.10 it can be shown, that a Bayesian network correctly describes the modeled system only if each node is conditionally independent of its parent nodes. This is an important constraint of BN construction. One has to select such parents for every node that this condition holds. Therefore directed circles are to be avoided. BN can efficiently describe especially sparse systems, where each variable influences only a limited subset of other variables, independently of the size of the system. In such case the system model can be significantly compressed compared to the full joint distribution. In contrast, a fully connected network would include the same amount of data as the full joint distribution.

It can be advantageous to construct a network iteratively, starting ideally with the first order parent(s), that become(s) the root(s) of the network. Then, the variables influencing the first order parents can be added and this process continued until the childless nodes are reached. It can be helpful to model the strongest factors first and extend the network later if needed. One can use a table with the variable fixtures to note the strength and the direction of dependency between all variables.

There are always different possibilities to (correctly) model a given system. Following of the above rules can result in different network topologies depending mainly on the initial first order parents' choice. The ideal result is represented by a sparse system with persistent causal topology. However, in some cases, a strictly causal model would cause a need for specifying additional dependencies between otherwise independent variables. In such cases it is of advantage to choose a so-called diagnostic approach. Its disadvantage is the fact that the conditional relationships usually require uncommon way of thinking. Both approaches can be specified as follows:

Causal model reflects the "more intuitive" causal direction of relationships. Some (hidden) cause produce a certain (observed) effect (e.g. change of the environment).

Diagnostic model makes use of the anti-causal relationships, which lead from (observed) effects to (hidden) cause: If certain effect was observed, there must be a certain cause.

It is possible to show both approaches to be correct and mathematically equivalent [RN03]. Although the causal modeling should be preferred, it depends on the system designer which model he implements. Sometimes it is not trivial to determine if a variable is a cause or effect (symptom). In practice, the resulting topology can be a mixture of both models.

Figure 3.4 shows an example of both alternatives. The operation of the alternator causes the battery to have a defined status as depicted in part a). If the state of the battery is observed (e.g. because empty) there must be something wrong with the alternator.

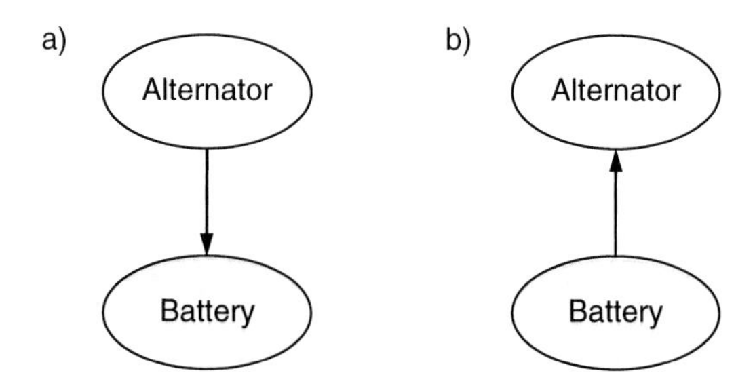

Figure 3.4: a) Causal and b) Diagnostic model example

3.3.2 Conditional probabilities

Conditional probabilities define the kind and the strength of relationships in a network. They are usually based on historical knowledge about how often a certain state occurred given fixed parent node's condition. Depending on the type of used variables, the conditional probabilities are represented either by probability tables or density functions (s. section 3.2). The size of CPT of node X depends on its parents Y_i and on the number of their possible values n_{y_i} and on the number of possible values n_x of the variable X itself:

$$N = n_x \prod_{\forall i} n_{y_i} \tag{3.16}$$

where N is the CPT size in cells. For parentless nodes $N = n_x$. Therefore, a high number of direct parents results in exponential boom of the CPT cells. This can be avoided by node grouping in hierarchical network structure taking the conditional (in)dependencies of individual nodes into account. If required, additional (synthetic) nodes can be inserted in the network.

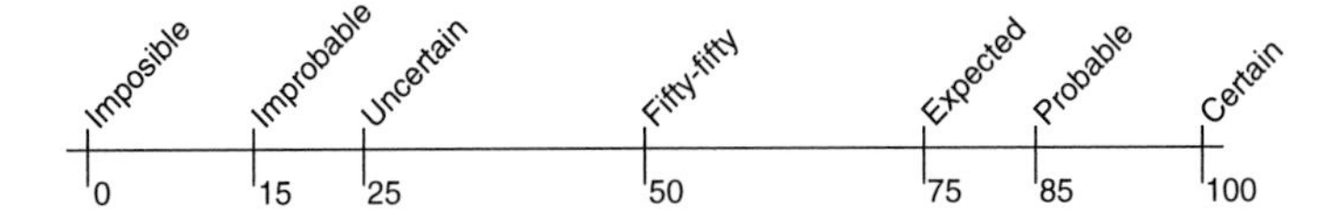

Figure 3.5: Numerical and verbal probability scale

In order to determine the probability distributions one can generally use the following sources of knowledge:

- Manufacturer's specifications
- Own measurements/ statistics
- Expert knowledge
- (Supervised) learning

Ideally, the manufacturer of the used system (component) provides his own statistics which suites for the desired purpose. If this is not the case, the optimal way is to generate own statistics according to the defined problem. Own statistics can be based on experimental measurements, as shown in chapter 5. In some cases, there is no statistics available and it is not possible to create one's own. In this case one may ask one or more experts for their opinion based on their experience. Instead of providing numerical probabilities, it can be easier to get the according verbal descriptions. Therefore, a so-called probability scale was invented [RW99], which can be used to transform verbal descriptions into numeric probabilities and vice versa (s. figure 3.5). The opinions of more experts can be combined with probability aggregation methods [XxHS07].

Given a network structure and a labeled representative data set, it is possible to "learn" conditional probabilities "automatically". This method is generally called parameter learning. It might be helpful especially while exploring new domains, where neither statistics nor expert knowledge is available. Learning in Bayesian networks is a huge research area, which goes into another direction compared to the focus of this thesis. Depending on the particular problem the appropriate method can be chosen from a variety of online or offline, supervised or unsupervised, static or dynamic and further methods described in literature [HGC94, Hec95, OP00, Mur02, For02, RN03, Rüd03, TGW+05].

3.4 Applied probabilistic inference

Probabilistic inference provides mechanisms to compute the resulting (posterior) probability distribution for a set of query variables, given a set of evidence variables (observed events). The non-evidence variables are also called hidden variables. In the graphical representation the hidden nodes are clear and the shaded nodes are observed. Inference algorithms are principally based on Bayesian theorem (equation 3.5) and the chain rule (equation 3.9). For small and middle sized networks, exact algorithms can be adopted that directly make use of probabilistic rules and apply them in optimal order (e.g. variable elimination). For very large networks such computations can become intractable (s. section 3.3.2) and therefore approximate methods based on sampling algorithms are beneficial. Details about possible implementations of inference algorithms can be found in literature [Pea88, Mur02, RN03, DKP+06]. Basically, the approaches make different trade-offs between accuracy, speed, generality, etc.

Basic principles of probabilistic inference in Bayesian networks are illustrated in the following by means of a basic example. All variables of the example network are implemented with two possible states. Continuous values are accordingly discretized. The used probability distributions are represented by sample values. The system is depicted in figure 3.6 and consists of three nodes with the following denotation:

Track Quality represents a measure of the quality of object's following (tracking) by an environment sensor.

Dropouts The number of missing detections, where a certain sensor object is lost and then detected again, in a defined time window.

Tracking time The time period since a certain object was initialized.

We assume the system is equipped with a "watchdog" algorithm observing independently the number of dropouts and the tracking duration for tracked objects. The query variable is the track quality.

In this network, the causal factors are defined as follows: Both the tracking duration as well as the dropout rate are directly influenced by the track quality. This is a typical example of causal modeling, where a hidden property causes a certain observable effect (cp. figure 3.4).

The probability distribution of the two child nodes is defined conditioned on particular values of the parent node. Hence, given a certain value of

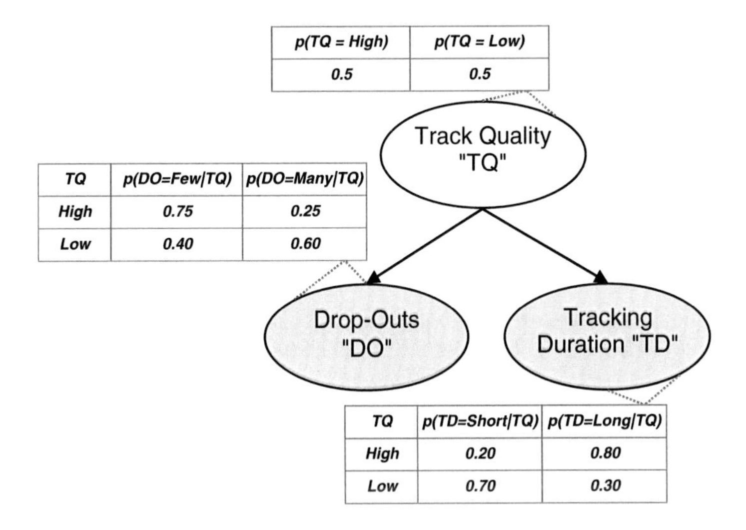

Figure 3.6: Example BN model for estimation of tracking quality

track quality, the probability of entire values of dropout rate $P(DO|TQ)$ and tracking duration $P(TD|TQ)$ is specified in the tables. The a priori probability of tracking quality $P(TQ)$ is defined uniformly.

Examples

Once the network was topologically constructed and the conditional probability distributions (tables) specified, the network can be used to process queries. Thereby, observed evidences can be incorporated as constraints. Two example questions and the according detailed answers based on this network (s. figure 3.6) are provided in the following:

Question 1: Suppose high tracking quality of an object, what is the probability that this object is being detected continuously with at most few dropouts?

Answer 1: The question can be expressed mathematically as follows:
$P(DO = Few|TQ = High) = ?$ This probability can be determined directly from the definition of the probability distribution of the dropout node:
$P(DO = Few|TQ = High) = 0.75.$ □

Question 2: What is the probability of a high tracking quality of an object under the adoption of a long tracking duration?

Answer 2: In terms of mathematics this question can be expressed as: $P(TQ = High|TD = Long) = ?$

Initially, one makes use of the generalized Bayesian theorem (3.5):

$$P(S|K) = \sum_{\forall F} P(S,F|K) = \sum_{\forall F} \frac{P(S,K,F)}{P(K)} \tag{3.17}$$

Dropouts are a free variable in this case, because no evidence about this variable is known. Hence, inserting in the above equation results in:

$$P(TQ = High|TD = Long) = \sum_{DO} \frac{P(TQ = High, TD = Long, DO)}{P(TD = Long)} \tag{3.18}$$

Then the chain rule (3.9) will be used whereby

$$X_1, ..., X_n = TQ, TD, DO \tag{3.19}$$

to transform the summand in the numerator of equation 3.18 as follows:

$$P(TQ, TD, DO) = P(TQ) \cdot P(TD|TQ) \cdot P(DO|TQ) \tag{3.20}$$

Finally, the equation 3.20 can be substituted in equation 3.18 and the numerical result can be obtained by inserting the numerical values specified in figure 3.6 in the resulting equation:

$$P(TQ = High|TD = Long) = \\ = \sum_{DO} \frac{P(TQ = High)P(TD = Long|TQ = High)P(DO|TQ = High)}{P(TD = Long)} = 0.727 \tag{3.21}$$

Verbally, this result could be translated as "expected" (s. figure 3.5). □

Chapter 4

Integrated probabilistic approach to environmental perception

The environmental perception of advanced driver assistance systems can be processed by a class of Bayesian estimators (s. section 2.2). The Kalman filter (KF) is probably the most famous member of this class. The filtering result is computed in the easiest case by applying several matrix operations (s. section 2.2.2). In this way one gets the optimal state estimation and its uncertainty represented by its distribution (KF: Gaussian). Many possible enhancements have been published (s. section 2.2.2) since Rudolf Kálmán introduced his method [Kal60]. They focus on the limiting assumptions of the original approach, trying to deal with nonlinear systems and/or non-Gaussian distributions. Nevertheless, the distribution of the estimated state vector components remains the only quality measure of the environmental perception (s. section 2.3.2). It is a helpful value, but not always representative and not sufficient for comprehensive analysis. It does not take further available aspects in account; e.g. existence probability, tracking quality, association success and the previous experience in general. Moreover, the only supported object's behavior prediction is based purely on the used dynamic model which is not able to incorporate any a priori or additional knowledge (e.g. lane resp. road parameters).

Instead of deriving a particular solution especially for one application, a consistent generalized approach to environmental perception with the ability to incorporate additional a priori information is desired (s. section 1.3). For this purpose the probabilistic point of view and the representation of the filtering algorithm as a recursive Bayesian estimator is adopted, which can be generally concerned as a special case of a class called Bayesian networks, a generic denomination for a largely applied class of probabilistic graphical models (s. section 3.2).

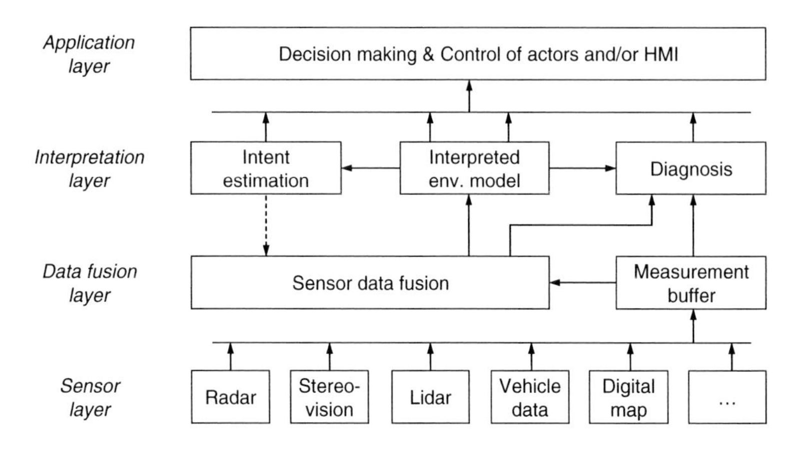

Figure 4.1: Environment perception processing layers

4.1 General strategy

The proposed environment perception system can be subdivided into several processing layers as depicted in figure 4.1. The first level is formed by a sensor network. In this layer the sensor measurements are collected, preprocessed and transmitted asynchronously over the according bus to the measurement buffer of the data fusion layer. The measurement data buffer ensures that the asynchronous sensor data are processed in a correct order.

In the data fusion layer an event controlled processing of the sensor data is done. Initially, the used sensor specific communication protocol is mapped on a generic sensor data structure and provided to the fusion core. Depending on its configuration, different fusion algorithms can be used and evaluated (e.g. measurement or track-to-track approach). Based on the sensor data a virtual environment model is constructed. In this model the complete available information about the host vehicle and its environment is stored (s. section 4.2). Within the framework of this thesis a probabilistic representation of a measurement data fusion and the novel track-to-track level fusion implementation by a Bayesian network is introduced and evaluated. The latter approach makes use of the sensor internal build-in tracking with the objective to minimize the total perception system latency and thus to maximize its effective accuracy.

In the proposed interpretation layer the objects of the environmental model are sorted in a predefined way according to their qualitative pose in the environment relatively to the host vehicle. As a result, a basic abstract description of the scenario is produced and made available to other perception modules and assistance applications (s. section 4.2.6).

The environment perception quality of the environment objects is estimated by the diagnosis module. It observes the raw sensor data as they arrive in the measurement buffer in parallel to the interpreted environment model and the data fusion internal variables. Based on this information and previous experience, a confidence level can be supplemented to each object of the environment model (s. section 4.3). The main innovation of this module represents the consideration of different symptoms across the uncertainty domains including the state variables uncertainty, detection uncertainty as well as the association uncertainty. Moreover, a special method is developed and implemented to experimentally determine the state variables and the detection uncertainty characteristics in the entire field of view of individual sensors. The probabilistic representation of this module allows a natural coupling with the data fusion algorithms and assures mathematical consistency.

Maneuver level intent estimation of traffic participants bases on the interpreted environment model and predefined behavior patterns (s. section 4.4). The dashed arrow depicts its optional coupling with the sensor data fusion module to improve its prediction step. The contribution of this thesis includes the general integration of a probabilistic intent estimation algorithm as an extension of the classical low level prediction by a priori knowledge. Furthermore, a proof of this concept by means of a lane change example maneuver is implemented.

The resulting environmental perception mechanism with diagnosis and intent estimation ability covers the broad field from sensor measurements up to the high level intent estimation of traffic participants. Thereby, Bayesian networks and thus the probability serve as uniform and consistent denominator.

The interpreted environment model enriched by diagnosis and intent estimation data forms a common interface for entire driver assistance applications included in the application layer, where a decision concerning the next action(s) is made and the actors and/or the HMI is controlled accordingly (e.g. giving a hint or intervene in the vehicle dynamics). In case of more assistance applications used simultaneously, a consistent functionality of all of them is ensured, because all applications take their decisions on basis of a common information base. This approach was successfully approved by means of supplying input data for the "Integrated Lateral Assistance" application described in chapter 5.

In the following the essential modules of this approach are presented in detail in accordance to [JFS+09].

4.2 Sensor data fusion

Sensor data fusion (SDF) represents the heart of multi-sensorial environmental perception (s. figure 4.1, data fusion layer). The input for the data fusion layer is collected from different sources (e.g. host vehicle, its surrounding area) by multiple sensors (s. figure 4.1, sensor layer).

The storage of the sensor data in the measurement buffer (s. figure 4.3) is necessary, because of their potentially big amounts arriving simultaneously. Furthermore the measuring time does not necessarily agree with the reception time of the data. The data records should be processed in the order of their measurement times and not in the order of their reception. Therefore during the read-out of the oldest data record from the buffer, it is tested, whether a data record with older measurement time is expected. In such case the fusion cannot be continued until this data set arrives. On the other hand, unnecessary idle times are to be avoided e.g. by defined timeouts. Instant data fusion according to the reception time would require the subsequent incorporation of data sets with older measuring times [KE08]. This complex problem is thereby avoided.

The data fusion layer gathers all buffered sensor data and fuses it into an environment model (EM) which serves as an image of the current "real" environment. The EM in turn serves as input to the other perception modules. Figure 4.2 shows a general SDF architecture and the appropriate data flow according to [Stü04].

The estimated state variables of all detected objects are stored in the EM (figure 4.2, top left). When a new sensor measurement set is delivered (figure 4.2, top right), the EM is predicted to the sensor measurements' time using defined dynamic model(s). The received sensor data contains the measured characteristics of the observed objects. The expected characteristics from the temporal prediction of the objects state are to be generated before the association step can pass. The association step compares

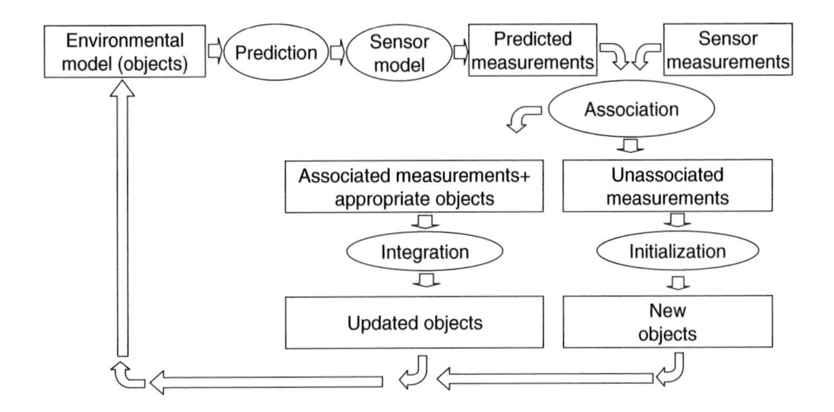

Figure 4.2: Sensor data fusion – general data flow

the measured and the predicted characteristics. In case of successful association, the measured characteristics are assigned to an object of EM and the state estimate of this object is updated with the associated measurements. The unassociated measurements are used for initialization of new objects. After the update step is finished, the validity of all objects in the EM is checked. The environment objects that have not been updated for a certain specified time are removed from the EM. Finally, the updated environment model can be released and further interpreted.

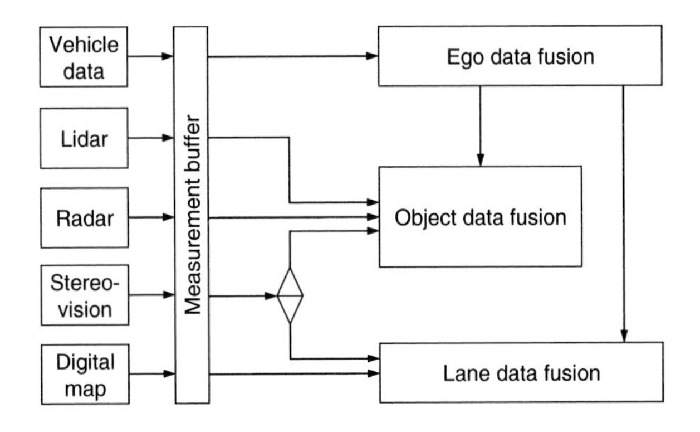

Figure 4.3: Sensor data fusion – internal architecture

Besides the fusion of the "common" environment objects the movement estimation of the host vehicle (based on on-board odometry sensors) is of significant importance. Environmental sensors deliver in general relative data. Therefore the host vehicle motion must be tracked precisely (s. figure 4.4). The same principle is valid for the lane course estimation. The lane is treated as special environment object. The information about lane course resp. lane markings is computed from the according sensors and the host vehicle movement estimation. The internal structure of sensor data fusion including the usage of individual sensors by the fusion modules is depicted in figure 4.3.

Within the described approach most of the common measurement data fusion algorithms can be used as described in section 2.2.2. From the probabilistic perspective, these methods can be seen by means of a so-called recursive Bayesian estimator, which can be represented by a dynamic Bayesian network as introduced in section 3.2.4. This is not a simple alternative, it allows one to work with more general models and to extend the restricted topology of the filtering algorithm as demonstrated in section 4.2.4.

As stated in the introduction to this chapter, this thesis generally adopts the probabilistic point of view using Bayesian networks.

4.2.1 Environmental objects modeling

The data from environmental sensors are processed in form of measurement vectors. In the measurement vector, ideally only real measurements should be considered and not derived values. Thus, e.g. velocity measurements from time-of-flight based sensor systems should be omitted. In practice, the contents of a measurement vector are determined for each sensor individually, based on its internal processing frequency and further parameters. This ensures optimal sensor data processing without any information loss and without providing any kind of bias. Table 4.1 summarizes the sensor types used in the experimental vehicle (cp. chapter 5) and the according measurement vectors.

The data fusion principle was described in section 4.2 and the internal architecture is depicted in figure 4.3. The object data fusion determines the spatial, geometrical and dynamical features of objects in the host vehicle surroundings by combining of measured data of different sensor systems. As a result, a fused object list that contains all detected objects in the

Sensor	Measurement vector $\mathbf{z}$
Radar	$[d_x\ d_y\ v_{rad}]^T$
Laser	$[d_x\ d_y\ w\ l]^T$
Stereo camera	$[d_x\ d_y\ w\ h\ l]^T$

Table 4.1: Typical measurement vectors of different sensor types

vehicle surroundings is produced. The object list generated by the object data fusion serves as essential input for further processing. The fused object list contains generally objects, which are in the field of view of the used environmental sensors.

The state of each detected environmental object is modeled by its state vector. It bases on the underlying dynamic model with constant acceleration [MW01, BP99], which is used for prediction of objects' movements and contains the following components:

$$\mathbf{x} = [d_x \quad d_y \quad v_x \quad v_y \quad a_x \quad a_y \quad w \quad h \quad l]^T \tag{4.1}$$

where

d_x	[m]	relative coordinate of an environment object to the center of the host vehicle's front bumper, parallel to driving direction, positive to the front
d_y	[m]	relative coordinate of an environment object to the center of the host vehicle's front bumper, perpendicular to the driving direction, positive to the left
v_x, v_y	[m/s]	absolute components of velocity (over ground) of an environment object in direction of d_x and d_y
a_x, a_y	[m/s²]	absolute acceleration (over ground) of an environment object in d_x and d_y direction
w	[m]	width of an environmental object
h	[m]	height of an environmental object
l	[m]	length of an environmental object

4.2.2 Host vehicle modeling

The ego data fusion module (s. figure 4.3) combines the host vehicle data and provides them for other functional modules. It processes the data from vehicle on-board odometry sensors. Thereby constant acceleration model of vehicle dynamics is used as in the case of common environment objects.

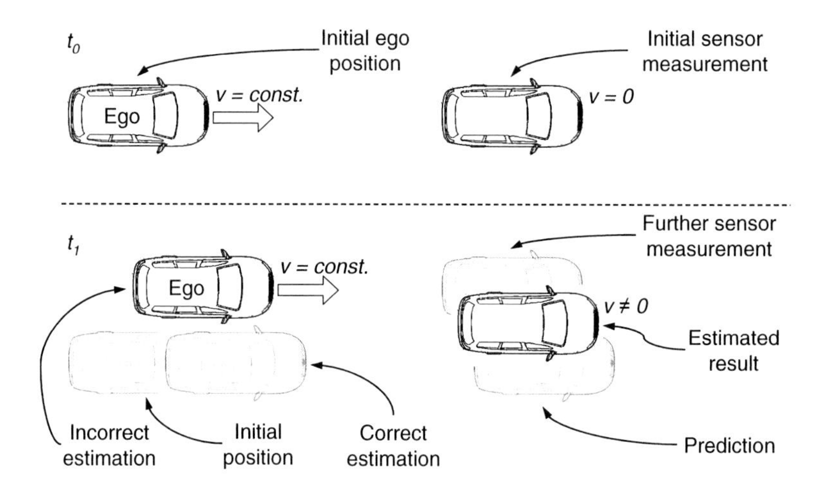

Figure 4.4: Consequence of host vehicle movement estimation error for object detection

Initially, the data of the on-board odometry sensors for the ego fusion module is preprocessed, producing the following pseudo-measurement vector.

$$\mathbf{z} = [v_x \quad \dot{\psi} \quad v_y]^T \tag{4.2}$$

v_x	[m/s]	longitudinal velocity
$\dot{\psi}$	[rad/s]	yaw rate
v_y	[m/s]	lateral velocity

The on-board lateral acceleration sensor can be affected by the gravitation force, which can in certain situations (e.g. vehicle on a lateral slope) deliver highly inaccurate values, leading to further error propagation to other states. Therefore, instead of using directly the measurements from the lateral acceleration sensor a composed measurement of the lateral velocity v_y is used as measurement input.

The composed measurement signal for the lateral velocity is obtained using a linear tire model [Ise06]. The lateral velocity v_y can be calculated from the longitudinal velocity v_x, the yaw rate $\dot{\psi}$ and the lateral acceleration a_y as follows:

$$v_y = \dot{\psi}\frac{w_b}{2} - v_x k_{a_y} a_y \tag{4.3}$$

where w_b is the wheel base and k_{a_y} a design constant of the linear tire model. The resulting state vector of the host vehicle is modeled as follows:

$$\mathbf{x} = [e_x \quad e_y \quad v_x \quad v_y \quad a_x \quad a_y \quad \psi \quad \dot{\psi}]^T \tag{4.4}$$

e_x	[m]	global coordinate, 0 m at the start, orientation results from vehicle movement immediately after the start
e_y	[m]	global coordinate orthogonal to e_x, 0 m during the vehicle start, orientation results from vehicle movement immediately after the start
v_x	[m/s]	velocity in longitudinal direction
v_y	[m/s]	velocity in lateral direction, positive to the left
a_x	[m/s²]	longitudinal acceleration, acceleration along vehicle orientation
a_y	[m/s²]	lateral acceleration, acceleration perpendicular to the vehicle direction, positive to the left
ψ	[rad]	yaw angle, angle between global and vehicle coordinate system, positive yaw angle – vehicle rotated anticlockwise
$\dot{\psi}$	[rad/s]	yaw rate, yaw angle change

The lateral acceleration component of the state vector is omitted by the filter algorithm. Its input value is processed separately with respect to its estimated offset and forwarded directly into the state vector.

Figure 4.4 illustrates the importance of precise modeling of the host vehicle dynamics and the consequence of an estimation error for object detection. Assume a stationary object was detected at time t_0 by environmental sensors direct in front of the ego vehicle (figure 4.4, top). Due to an imprecise movement estimation of the host vehicle at time t_1 it drifts globally sidewards (figure 4.4, bottom). In the meantime, the sensor system detects the target object again directly in front of the host vehicle. Because of the corrupted prediction, the filtering result is wrongly somewhere between the predicted and the measured position. Furthermore, a movement of the stationary target vehicle is wrongly assumed!

4.2.3 Driving lane modeling

A common and well-approved horizontal model in the domain of road design, especially of highways, is a so-called clothoid [DZ86]. Clothoid is represented by a curve, whose curvature κ is proportional to its length:

$$\kappa(l) = \kappa_0 + \dot{\kappa} l \tag{4.5}$$

where κ_0 stands for the initial curvature and $\dot{\kappa}$ is the derivation of the curvature, which describes how the curvature variates.

The measurement vector of a driving lane is defined as follows:

$$\mathbf{z} = \begin{bmatrix} \kappa_0 & \dot{\kappa} & \alpha & d_l & d_r \end{bmatrix}^T \tag{4.6}$$

κ_0	[1/m]	initial horizontal curvature
$\dot{\kappa}$	[1/m²]	horizontal curvature variation
α	[rad]	relative angle between the vehicle longitudinal axis and the lane axis
d_l, d_r	[m]	relative distance to the left and to the right lane markings

Hence, the skeletal line (axis) of a lane model has a shape of a clothoid in horizontal direction. Under certain assumptions, the following polynomial approximation of third degree can be used for the calculation of a clothoidal point $y(x)$ in the longitudinal distance of x.

$$y(x) = \frac{1}{6}\dot{\kappa}x^3 + \frac{1}{2}\kappa_0 x^2 + \alpha x + y_0 \tag{4.7}$$

The interpretation of the variables is in accordance to equation 4.6.

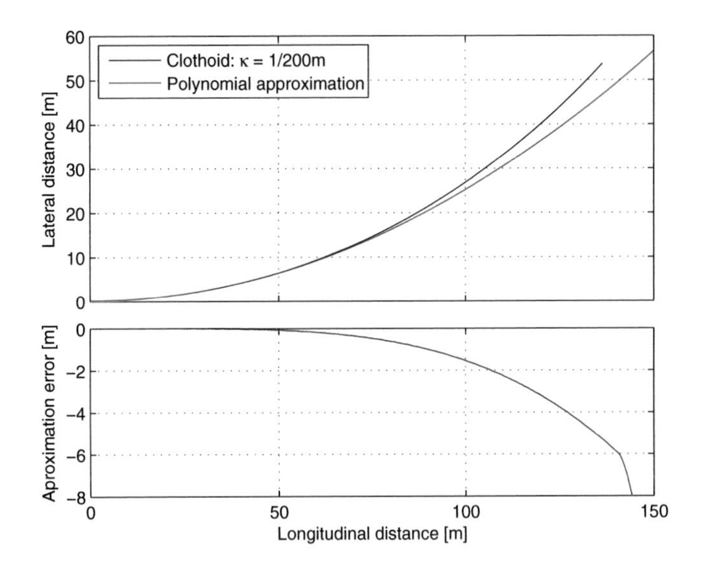

Figure 4.5: Polynomial approximation of a clothoid with radius of 200m (top) and the according approximation error (bottom)

The above approximation is only possible for lanes with a limited curvature, as they are for example on highways. This is demonstrated by an example visualization in figure 4.5, which compares a clothoid to this approximation (top), and depicts the according approximation error (bottom). Even in the case of a strong curvature on the limit of today's sensor specifications, the approximation error in the detection area (about 50m) does not exceed several centimeters.

The lane course prediction is based on constant curvature model. In case of temporally insufficient quality of lane markings or their occlusion, the last plausible state of the driving lane can be further updated based on the host vehicle movement. The information from digital map is intended to be used for long range support of assistance applications (e.g. velocity adaptation in front of a curve).

4.2.4 Recursive Bayesian estimator

A recursive Bayesian estimator represented by means of dynamic Bayesian network provides an alternative advantageous description of classical filtering techniques as announced in section 4.2.

Figure 4.6 shows basic structure of Bayesian estimator represented by a dynamic Bayesian network (s. section 3.2.4). The standard convention used e.g. by Murphy in [Mur02] is followed, and shading means a node is observed; clear nodes are hidden. $\mathbf{X}_t$ corresponds to the estimated state of the target object at time t; $\mathbf{z}_t$ is an arbitrary sensor measurement at time t assigned to the evidence variable $\mathbf{Z}_t$. Thereby, the Markov property for $\mathbf{X}_t$ and the independence of the measurements $\mathbf{z}_t$ are assumed. Thus, $\mathbf{P}(\mathbf{X}_{t+1}|\mathbf{X}_t)$ represents the state evolution (dynamic) model. The sensor model definition corresponds to $\mathbf{P}(\mathbf{Z}_{t+1}|\mathbf{X}_{t+1})$. In the following the derivation of both the prediction and the update step is demonstrated.

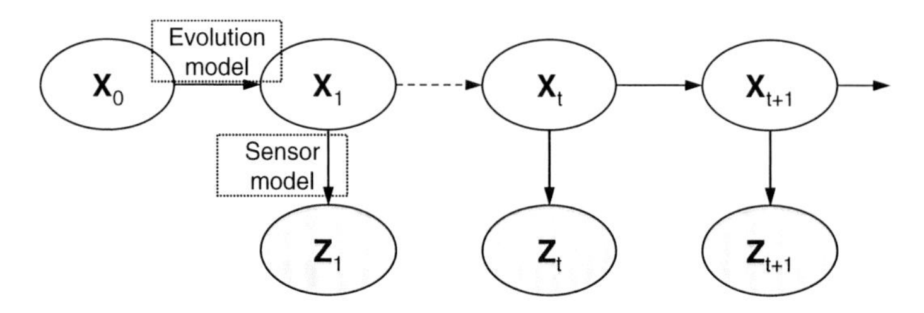

Figure 4.6: Recursive filter represented by a dynamic Bayesian network

The updated state $\mathbf{X}_{t+1}$ conditioned on the measurement set including the latest measurement is defined by:

$$\mathbf{P}(\mathbf{X}_{t+1}|\mathbf{z}_{1:t+1}) = f(\mathbf{z}_{t+1}, \mathbf{P}(\mathbf{X}_t|\mathbf{z}_{1:t})) = \mathbf{P}(\mathbf{X}_{t+1}|\mathbf{z}_{1:t}, \mathbf{z}_{t+1}) \tag{4.8}$$

Then, by applying of the conditional Bayes theorem (s. equation 3.12) one gets:

$$\mathbf{P}(\mathbf{X}_{t+1}|\mathbf{z}_{1:t+1}) = \frac{\mathbf{P}(\mathbf{z}_{t+1}|\mathbf{X}_{t+1}, \mathbf{z}_{1:t}) \cdot \mathbf{P}(\mathbf{X}_{t+1}, \mathbf{z}_{1:t})}{P(\mathbf{z}_{t+1}|\mathbf{z}_{1:t})} \tag{4.9}$$

And the latter incorporation of the Markov property results in:

$$\mathbf{P}(\mathbf{X}_{t+1}|\mathbf{z}_{1:t+1}) = \frac{\mathbf{P}(\mathbf{z}_{t+1}|\mathbf{X}_{t+1}) \cdot \mathbf{P}(\mathbf{X}_{t+1}, \mathbf{z}_{1:t})}{P(\mathbf{z}_{t+1}|\mathbf{z}_{1:t})} \tag{4.10}$$

One can recognize the first term in the numerator of 4.10 already mentioned above, namely $\mathbf{P}(\mathbf{z}_{t+1}|\mathbf{X}_{t+1})$. It can be derived directly from the definition of the sensor model. The denominator acts as a normalizing constant. The second term in the numerator represents the prediction of the next state. It can be expressed as integral (sum for discrete case) of the product of all probabilities of any complete set of mutually exclusive hypotheses of $\mathbf{x}_t$ and corresponding conditional probabilities as follows:

$$\mathbf{P}(\mathbf{X}_{t+1}|\mathbf{z}_{1:t+1}) = \frac{\mathbf{P}(\mathbf{z}_{t+1}|\mathbf{X}_{t+1}) \cdot \int\limits_{\mathbf{x}_t} \mathbf{P}(\mathbf{X}_{t+1}|\mathbf{x}_t, \mathbf{z}_{1:t}) \cdot P(\mathbf{x}_t|\mathbf{z}_{1:t})\, d\mathbf{x}_t}{P(\mathbf{z}_{t+1}|\mathbf{z}_{1:t})} \tag{4.11}$$

The resulting term can be simplified assuming that the Markov property is met:

$$\mathbf{P}(\mathbf{X}_{t+1}|\mathbf{z}_{1:t+1}) = \frac{\mathbf{P}(\mathbf{z}_{t+1}|\mathbf{X}_{t+1}) \cdot \int\limits_{\mathbf{x}_t} \mathbf{P}(\mathbf{X}_{t+1}|\mathbf{x}_t) \cdot P(\mathbf{x}_t|\mathbf{z}_{1:t})\, d\mathbf{x}_t}{P(\mathbf{z}_{t+1}|\mathbf{z}_{1:t})} \tag{4.12}$$

One can recognize the state evolution model $\mathbf{P}(\mathbf{X}_{t+1}|\mathbf{x}_t)$ and the current state distribution $P(\mathbf{x}_t|\mathbf{z}_{1:t})$ inside the integral. Hence, the filtering problem is theoretically solved.

Unfortunately, the equation 4.12 does not result in a closed solution in general. There are two general options how to deal with this issue:

On the one hand, there are some special cases with a closed solution. If the distribution of the state vector is Gaussian and additionally the evolution and the sensor model are linear Gaussian, then the result of equation 4.12 is Gaussian as well (cp. section 2.2.2). It can be shown, that the family of Gaussian distributions remains closed under the Bayesian network operations. Furthermore, if all state variables are discrete, constant time and memory amount is required to obtain the solution [RN03].

On the other hand, dynamic Bayesian networks are generally able to handle any arbitrary probability distributions. In this case approximative inference methods may need to be applied, if exact inference is not possible. In addition, dynamic Bayesian networks allow in contrast to classical techniques a fast network topology redesign, as demonstrated in section 4.2.5 and depicted in figure 4.7.

4.2.5 Track-to-Track data fusion

This section demonstrates a novel application and the extensibility of the dynamic Bayesian estimator introduced in section 4.2.4. Thus, a track-to-track data fusion approach for sensors with built-in internal tracking is derived.

Its main purpose is to prevent additional filtering delays of a pseudo measurement filter cascade and to take maximal use of sensor specific supplier's knowhow (s. section 2.2.2).

The bottom part of figure 4.7 depicts internal recursive Bayesian estimators of the example sensors A and B (cp. figure 4.6) with typically unknown supplier-specific tracking parameters. The notation is analog to figure 4.6.

The upper part of figure 4.7 will be further analyzed in this section. The fusion result is computed by a combination of the current sensor data set (from any sensor) and the associated tracked data output from the second sensor.

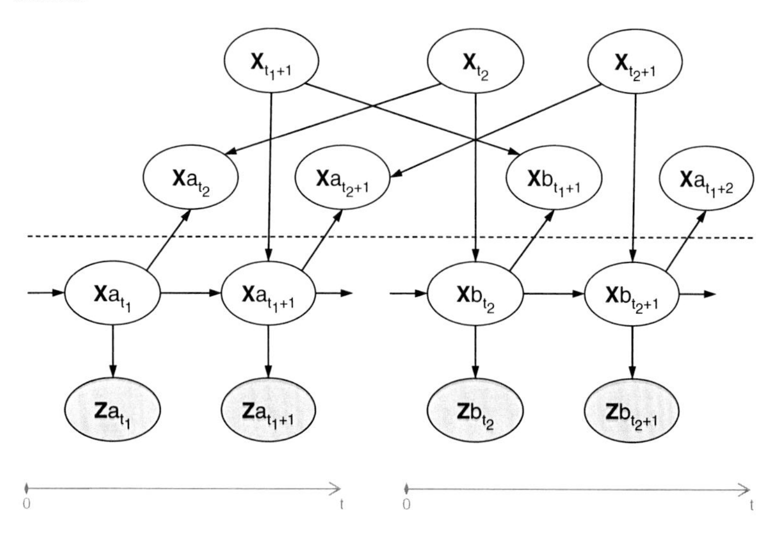

Figure 4.7: Track-to-Track approach modeled by a Bayesian network

The following sequence order is supposed in this example: In track-to-track fusion, the states to be fused must be synchronized, i.e. must have the same time stamp. In general, however, sensors output data at different cycle times with non-synchronized local clocks. Therefore, the prediction of all sensor data to the time of the last (currently fused) data set is needed. Thereby a successful association of this sensor data is assumed.

The system variables are defined as follows:

- Estimated result $\mathbf{X}$ with defined a priori distribution $\mathbf{P}(\mathbf{X})$
- Output $\mathbf{X}_A$ from sensor A with defined conditional distribution $\mathbf{P}(\mathbf{X}_A|\mathbf{X})$
- Output $\mathbf{X}_B$ from sensor B with defined distribution $\mathbf{P}(\mathbf{X}_B|\mathbf{X})$

$\mathbf{P}(\mathbf{X})$ is the a priori distribution of the estimated result. Due to no prior knowledge about $\mathbf{X}$ a uniform prior density is the best way to express the "complete ignorance" of $\mathbf{X}$ [Jay68]. This ensures all values to be equally likely. Notice that a theoretical uniform distribution cannot be normalized. In practice one always has some kind of prior knowledge about location and scale (e.g. sensor field of view), and in consequence the boundary parameters cannot vary over a truly infinite range.

In order to handle $\mathbf{P}(\mathbf{X}_A|\mathbf{X})$ and $\mathbf{P}(\mathbf{X}_B|\mathbf{X})$ one has to specify how both distributions over the sensor measurement depend on the value of $\mathbf{X}$, that means to specify the parameters of $\mathbf{P}(\mathbf{X}_A|\mathbf{X})$ and $\mathbf{P}(\mathbf{X}_B|\mathbf{X})$ as a function of $\mathbf{X}$. This corresponds to a definition of both sensor models. The data fusion is represented by $\mathbf{P}(\mathbf{X}|\mathbf{X}_A,\mathbf{X}_B)$. Assume sensor A measures a value of $\mathbf{a}$ and sensor B measures a value of $\mathbf{b}$. Hence, $\mathbf{X}_A = \mathbf{a}$ and $\mathbf{X}_B = \mathbf{b}$. This can be expanded using the Bayesian rule (equation 3.11) as follows:

$$\mathbf{P}(\mathbf{X}|\mathbf{a},\mathbf{b}) = \frac{\mathbf{P}(\mathbf{X},\mathbf{a},\mathbf{b})}{P(\mathbf{a},\mathbf{b})} \tag{4.13}$$

Then, the chain rule (equation 3.9) can be applied for the nominator:

$$\mathbf{P}(\mathbf{X}|\mathbf{a},\mathbf{b}) = \frac{\mathbf{P}(\mathbf{X}) \cdot \mathbf{P}(\mathbf{a}|\mathbf{X}) \cdot \mathbf{P}(\mathbf{b}|\mathbf{X})}{P(\mathbf{a},\mathbf{b})} \tag{4.14}$$

And the denominator can be expressed as integral (cp. equation 4.11):

$$\mathbf{P}(\mathbf{X}|\mathbf{a},\mathbf{b}) = \frac{\mathbf{P}(\mathbf{X}) \cdot \mathbf{P}(\mathbf{a}|\mathbf{X}) \cdot \mathbf{P}(\mathbf{b}|\mathbf{X})}{\int_{\mathbf{X}} \mathbf{P}(\mathbf{X},\mathbf{a},\mathbf{b})\, d\mathbf{X}} \tag{4.15}$$

Finally, the chain rule can be applied again, this time for the denominator:

$$\mathbf{P}(\mathbf{X}|\mathbf{a},\mathbf{b}) = \frac{\mathbf{P}(\mathbf{X}) \cdot \mathbf{P}(\mathbf{a}|\mathbf{X}) \cdot \mathbf{P}(\mathbf{b}|\mathbf{X})}{\int_{\mathbf{X}} [\mathbf{P}(\mathbf{X}) \cdot \mathbf{P}(\mathbf{a}|\mathbf{X}) \cdot \mathbf{P}(\mathbf{b}|\mathbf{X})]\, d\mathbf{X}} \tag{4.16}$$

$\mathbf{P}(\mathbf{X})$ is defined as uniform distribution and the integral in the denominator is constant. Hence an important relation about the proportionality of the sensor measurements can be expressed:

$$\mathbf{P}(\mathbf{X}|\mathbf{a},\mathbf{b}) = \alpha \cdot \mathbf{P}(\mathbf{a}|\mathbf{X}) \cdot \mathbf{P}(\mathbf{b}|\mathbf{X}) \tag{4.17}$$

where

$$\alpha = \frac{\mathbf{P}(\mathbf{X})}{\int\limits_{\mathbf{X}} [\mathbf{P}(\mathbf{X}) \cdot \mathbf{P}(\mathbf{a}|\mathbf{X}) \cdot \mathbf{P}(\mathbf{b}|\mathbf{X})]\, d\mathbf{X}}$$

Thus $1/\alpha$ can be treated as a normalization factor and the likelihood (unnormalized probability) of the fusion result can be calculated by multiplication of the fixed conditional probability distributions $\mathbf{P}(\mathbf{a}|\mathbf{X})$ and $\mathbf{P}(\mathbf{b}|\mathbf{X})$. Thereby, $\mathbf{P}(\mathbf{X}_A = \mathbf{a}|\mathbf{X})$ resp. $\mathbf{P}(\mathbf{X}_B = \mathbf{b}|\mathbf{X})$ can be obtained directly from the sensor models definition.

Since there is no forward propagation on the top system level (figure 4.7, top), arbitrary probability distributions could be used for Gaussian and non-Gaussian processes without the need for approximative inference methods. In this regard, this algorithm outperforms the algorithms introduced in section 2.2.2. The measurement correlation between the used sensors is treated in accordance with the correlation suppression method (section 2.2.2). The method works analog with more sensors. In order to enhance the data association robustness further methods to improve its performance can be used (e.g. association history).

4.2.6 Interpreted environment model

The interpreted environment model bases on state variables of the environment objects estimated by the sensor data fusion algorithm from sensor data stored in its internal environment model. It provides a customizable output interface for assistance applications, for the intent estimation module and for the diagnosis module as well.

The objective of the interpreted environment model is to efficiently express the relations between the host vehicle and selected environment objects of interest in the context of traffic. The presented model is considered mainly for highway applications. Individual environment objects, in particular the host vehicle, the other vehicles and the driving lanes, are integrally considered and valued regarding to their relations to each other. The result is a compact description of the traffic scenario with respect to the host vehicle.

It is represented by an abstract matrix. This matrix contains driving lane parameters from lane fusion, host vehicle data from ego fusion and a symbolic object matrix containing in this case at most seven selected environmental objects (s. figure 4.8).

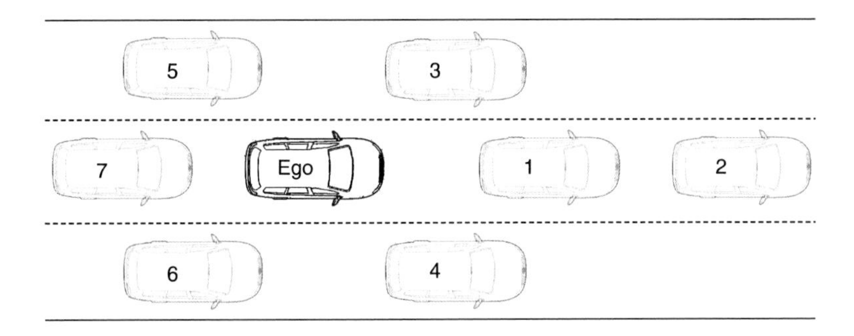

Figure 4.8: Symbolic object matrix as interpreted EM

In the following, the matrix qualifying criteria for environmental object are listed, as they have been specified for the experimental system (s. chapter 5):

1. *ACC target:* The nearest object in the front occupying the same lane as the host vehicle (ego). Between its left side and the left lane marking is not enough space for the host vehicle to pass safely.
2. *Hidden ACC target:* The second nearest potential object according to the definition of *ACC target.*
3. *Front left object:* The nearest object in front of the host vehicle in the left neighbor lane. This object can be located on the lane marking line or even partly in the same lane as the host vehicle (ego), but allowing enough space to pass. Otherwise it would be treated as *ACC target.*
4. *Front right object:* The nearest object in front of the host vehicle in the right neighbor lane, analog to *Front left object.*
5. *Rear left object:* The nearest object behind (or just besides) the host vehicle in the left neighbor lane.
6. *Rear right object:* The nearest object behind (or just besides) the host vehicle in the right neighbor lane.
7. *Rear center object:* The nearest object behind the host vehicle in the same lane as the host vehicle.

The lane assignment of the objects in near range (defined according to the lane detection range) bases on the clothoidal lane markings model

(s. section 4.2.3). Beyond that and in the rear area, it bases on estimated trajectories of the host vehicle and the detected vehicles.

4.3 Self-diagnosis

Today's advanced driver assistance applications directly process the input quantities of data from environment sensors. Thereby, solely general plausibility constraints based on generic thresholds are considered (low level diagnosis, s. section 2.3.1). Application designers rarely design the algorithms to handle *estimated* data. While the support degree of modern assistance systems is continuously increasing, a mechanism to ensure the reliability of the environmental perception data becomes necessary. Therefore, a novel high level diagnosis approach is presented in this chapter.

4.3.1 Basic idea

The key idea underlaying this diagnosis approach is to respect the fact the sensors cannot detect all possible targets under all possible circumstances (e.g. misalignment, bad weather conditions, pollution or aging). The objective is to recognize such situations and to initiate an appropriate system reaction if needed.

Thus, the environment perception algorithm must implement an internal mechanism to decide, how "trustworthy" is the information output transmitted to the assistance applications. Therefore, the input sensor data streams and the internal tracking algorithm flow is monitored on plausibility and consistency. This process bases among others on prior detailed knowledge of the used hardware (especially sensors).

Additionally, a long term diagnosis method is proposed, which monitors the balance of the network and which is able to identify potential long term issues.

4.3.2 General approach

Based on the observable symptoms of sensor data quality the confidence level of the environment objects can be estimated, assigned and further

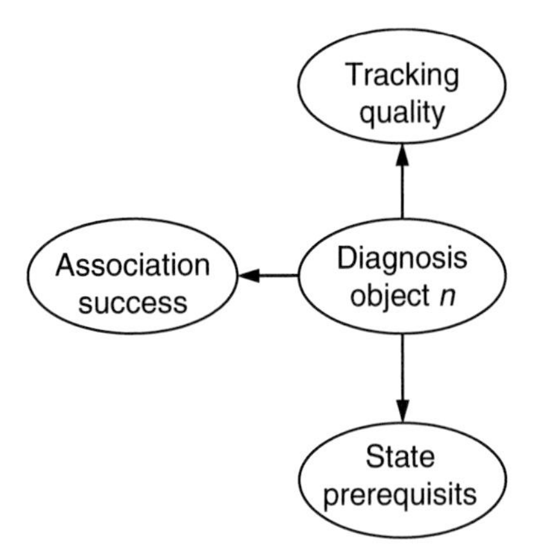

Figure 4.9: General diagnostic network of object n with main influence domains

communicated. [JMFN09] The main identified domains of symptoms are listed in the following:

- State prerequisites
- Sensor tracking
- Association success

These domains correspond exactly to the sources of uncertainty enumerated in section 2.3.2. In particular, the "State prerequisites" domain addresses the uncertainty of the state variables, the "Sensor tracking" domain addresses the detection uncertainty and the "Association success" domain addresses the association uncertainty.

Figure 4.9 shows the general topology of a Bayesian network, on which base the confidence level of an environmental object can be determined. The network causally modeled: a hidden property of interest (diagnosis) causes certain effects to be generated. Later on, these domains will be further divided. In the next sections the meaning and the way of modeling of the listed domains is explained in detail.

State prerequisites

The state prerequisites node represents a measure of how well the current state of an environment object can be captured by a specific sensor, based on previous knowledge (experience). It can be seen as the a priori state complexity. It can be modeled as a direct consequence of the hidden diagnosis node.

The state prerequisites node combines hierarchically further characteristic state vector components (s. figure 4.10). Independently of any other evidence, the object state estimate is observed in any case and can be therefore modeled by a set of evidence nodes. State prerequisites correspond to experience-based a priori variances of a sensor.

The purpose of this subnetwork (s. e.g. section 4.3.4) is related the state uncertainties computation, which is commonly represented by a covariance matrix (s. section 2.3.2). Unfortunately, todays automotive sensors provide no on-line (dynamic) uncertainty indication in addition to the state estimates in general. Usually, there are solely rough specifications concerning the general sensor accuracy provided. Furthermore, the expressiveness of theoretical models is always limited.

Because of its fundamental significance a special experimental procedure was developed to determine the sensor characteristics in detail. For this procedure, the host and an additional target vehicle is required. The host vehicle carries the sensor(s) to be evaluated and the target vehicle is equipped with a reference positioning system – a high accuracy DGNSS[1] receiver unit combined with an inertial system (s. appendix A). In order to obtain the sensor characteristics equidistantly over its complete field of view, the target vehicle is controlled automatically [JN09], scanning a defined pattern through the sensor field of view (s. figure 4.11)). Thereby, the reference values of the target vehicle dynamical features (e.g. relative position and velocity) are determined by a reference positioning system. The trajectory planning of the control algorithm can be processed by the reference positioning system as well and logged for later analysis. The host vehicle is stationary and does not move during the measurement. The according sensor values are recorded for every position in the entire sensor field of view in selected density according to the chosen trajectory based on the scenario of interest (e.g. longitudinal traffic, figure 4.11).

[1] Differential Global Navigation Satellite System

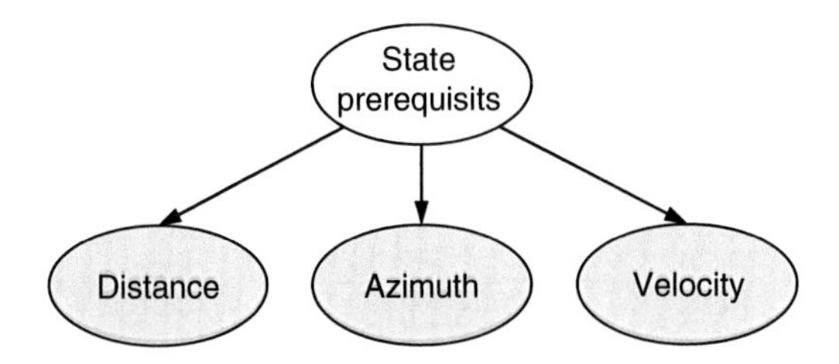

Figure 4.10: State prerequisites example subnetwork

The following basic data sets are to be determined for each sensor, based on the sensor data (host vehicle) and the target data (target vehicle):

- Sensor measurements $\mathbf{z}_s$ and dropouts d_s in the tested area of interest
- Reference measurements $\mathbf{z}_r$ of the geometrical and dynamical features of the target

The sampled area of interest should contain the entire field of view of the evaluated sensors. Afterwards, raw reference data are combined with geometrical features of the test object and the pose of the host vehicle to get the final reference data, which can be compared to the measured sensor data set.

Moreover, the following values of interest can be estimated by a direct analysis of the experimental data:

- Real field of view of a sensor assuming a measurement set $\mathbf{z}_{s_{1:N}}$ of N measurements corresponds to its hull. In case of a single beam sensor this hull is typically convex:

$$H_{convex}(\mathbf{z}_{s_{1:N}}) = conv\{\mathbf{z}_{s_1}, \mathbf{z}_{s_2} \dots \mathbf{z}_{s_N}\} \tag{4.18}$$

- Measurement error of i-th measurement of the evaluated sensor

$$\mathbf{\Delta}_i = \mathbf{z}_{s_i} - \mathbf{z}_{r_i} \tag{4.19}$$

The area of interest can be divided into more segments (e.g. Cartesian or radial grid layout). This step allows the incorporation of statistical methods, in particular the statistical estimation of the following values:

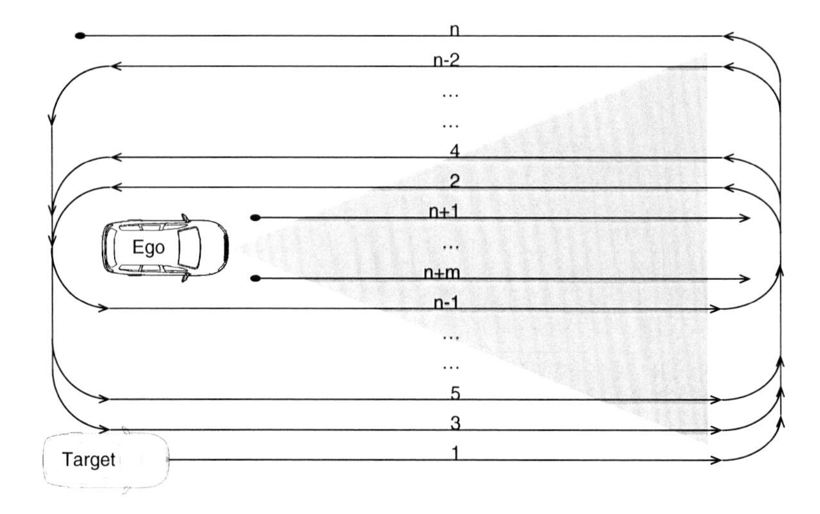

Figure 4.11: Scanning pattern to sample uniformly through the sensor field of view according to the "ice resurfacer" principle; n pattern sub-scans, m additional straight scans

- Mean error and standard deviation of all referenced features: The mean error $\overline{\mathbf{\Delta}}_A$ for a defined segment A with n measurements is defined by

$$\overline{\mathbf{\Delta}}_A = \frac{1}{n}\sum_{i=1}^{n} \mathbf{\Delta}_{A_i} \tag{4.20}$$

  and the sample standard deviation $\mathbf{s}_A$ of segment A:

$$\mathbf{s}_A = \left[\frac{1}{n-1}\sum_{i=1}^{n}(\mathbf{\Delta}_{A_i} - \overline{\mathbf{\Delta}}_A)^2\right]^{\frac{1}{2}} \tag{4.21}$$

- Existence probability: Probability that a detected object in the according segment A is a real object and not a sensor error consequence (false positive):

$$P_{ex} = \frac{N_{targets}}{N_{targets} + N_{ghosts}} \tag{4.22}$$

The mean $\overline{\mathbf{\Delta}}_A$ expresses the average deviation of the measurements in segment A from the desired value. Assumed a well-calibrated sensor and sufficient number of measurements, the determined means of all segments in the area of interest should be approximate zero. The according standard deviations describe how much the measurements spread around this value.

Tracking quality

The tracking quality represents a measure of the quality of object's following (tracking) by a defined sensor. It could be also referred to as track stability. This measure is further causally divided into two evidence nodes (s. figure 4.12). On the one hand the so-called dropouts variable is introduced, which represents the number of missing detections, where the object tracked by a sensor is lost and then detected again, in a defined time window. On the other hand the so-called tracking duration variable is used, which expresses the time period since the object was initialized. In fact, these variables are not directly observable. They result from data preprocessing done in the measurement buffer.

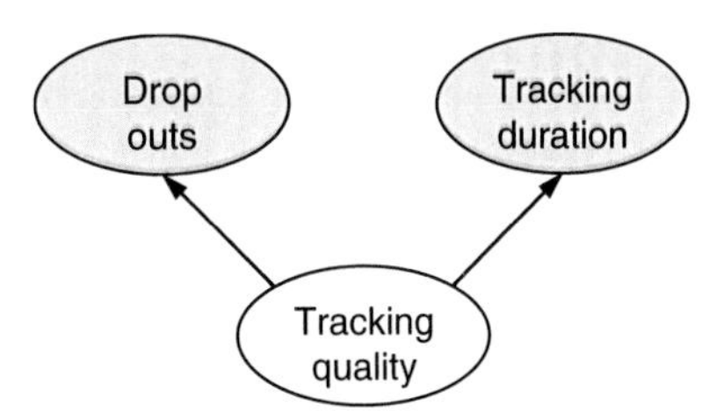

Figure 4.12: Tracking quality subnetwork

The parameters necessary for the determination of the according conditional probabilities are typically not provided by sensor suppliers. Therefore the parameters are to be determined experimentally based on the procedure described in section 4.3.2. Thereby, the general sensor performance especially in occurrence of dropouts is of interest, typically on the range-limits of the sensor.

The dependence between the tracking quality and the tracking duration can be determined from a set of referenced measurements $\mathbf{z}_{k:l} \neq \varnothing$, with $\mathbf{z}_{0:k-1} = \varnothing$. This is the case when a sensor measurement gets available after a defined break, e.g. when the target enters the field of view of the sensor. The set of measurements $\mathbf{z}_{k:l}$ can be further split into n sets according to the track stability progress:

$$\mathbf{z}_{k:l} = \left\{\mathbf{z}_{k:i_1}, \mathbf{z}_{k:i_2}, \ldots, \mathbf{z}_{k:i_n}\right\} \tag{4.23}$$

where $i_{1:n}$ are in accordance with given variance thresholds. This method can be applied to the whole measurement sequence to get reliable results valid in general.

In addition to the sensor measurements, the technique described in section 4.3.2 can be used to determine the dropouts as well. Based on this information, a general relation between the tracking quality/stability and the dropouts effect of various degree can be proposed. Therefore the threshold-based segmentation in subsets is required analog to equation 4.23 and propagated through the entire field of view of the sensor.

Association success

Successful measurement association between a sensor measurement and a tracked object expresses the nearness of these values to be below a defined threshold. Thus, the association success is modeled by a single variable (node) and not further split (s. figure 4.13).

Figure 4.13: Association success node

For the sake of measurements' association the expected values from the temporal prediction of the state of the available tracked objects is calculated. The association assigns the measured features of detected objects to the predicted features. A common decision criterion for the assignment is derived from the Mahalanobis distance (cp. section 2.3.2) between the measured and predicted object's features defined as:

$$d_t^2 = \mathbf{v}_t^T \mathbf{S}_t^{-1} \mathbf{v}_t \tag{4.24}$$

where

$$\mathbf{v}_t = \mathbf{z}_t - \hat{\mathbf{z}}_t \tag{4.25}$$

represents the current innovation and S_t the according covariance matrix (cp. section 2.3.2). The meaning of innovation is visualized in figure 4.4 as well.

The Mahalanobis distance provides a statistical distance, which considers among the state values their uncertainties as well. If the value of Mahalanobis distance is below a certain threshold[2], the measured features are assigned to an object. The association is processed for every sensor and every object in the according field of view. If an object is located in the field of view of more sensors, the measurements of all these sensors can be sequentially assigned.

Potential model deviation is signaled by a large innovation, because the prediction does not sufficiently describe the real object dynamics. Permanently increased innovation can point to a possible sensor misalignment as well. Both cases can result in inaccuracy and hence in an error-prone environment object's estimation.

Small Mahalanobis distance means, that the measured value lies close to the predicted one. In this case the state prediction matches (and therefore the model as well) with the actual trajectory in the state space and the model suitability is correspondingly higher than in the case of high Mahalanobis distance. For the environment object this means an increased probability of its accuracy.

Therefore the association success is used as an indicator of the environment model accuracy and correctness as a result.

Long term diagnosis

The effect of potential sensor malfunction can be gradual and indicated with an unpredictable and phased progression.

The static sensor characteristics based on the "State prerequisits" (s. section 4.3.2) represents the a priori knowledge about the sensor mean performance in its entire field of view. Therefore, the long-term mean contribution of dynamic domains of uncertainty ("Association success", "Tracking quality") should be theoretically neutral.

Long term diagnosis makes use of complementary adaptive characteristics additionally to the static sensor characteristics, which expresses the

[2] This threshold can be determined from the χ^2 distribution by defining a fixed confidence level [Pap01].

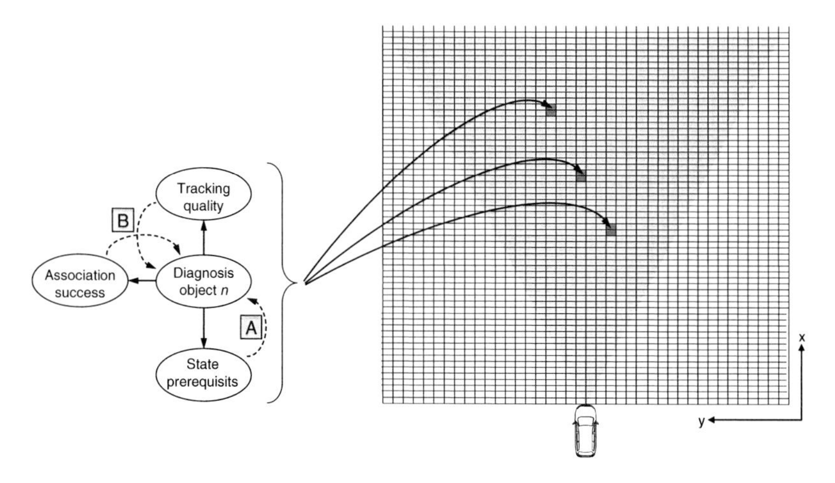

Figure 4.14: Long term diagnosis principle and the adaptive characteristics update based on the strength of dynamic influence

long-term deviation from the static values of the diagnosis network caused by the contribution from dynamic domains according to the original basic characteristics. Figure 4.14) depicts the static inference contribution A and the dynamic (deviance) contribution B.

The sum of both characteristics reflects the current general characteristics of the system. The result can be used as information source in the processing chain instead of the basic characteristics. Furthermore, certain trends of the adaptive characteristics can point to potential issues. Hence, the presence of the long term complementary characteristics can additionally simplify eventual offline fault diagnosis.

4.3.3 Applied principles

The environmental perception of a vehicle consists in general of two important parts. The first one is the relative detection of the relevant environmental objects (incl. the lane(s) of the roadway). The second is the estimate of the movement of the host vehicle (odometry). Similarly as in section 4.2 the approach focuses on "common" environmental objects. Based on this, the diagnosis of host vehicle and lane estimation can be derived. Unlike the section 4.2 discrete variables in combination with static Bayesian networks have been preferred. Hence, more freedom concerning the distributions (transitions) arises and an additional challenge concerning the eventual variable discretization is to be handled.

The diagnosis can be referred to as a measure of confidence of an object. The result of the diagnosis represents the probability, that given requirements of the application (set) are fulfilled. For an object n, it can be expressed by probabilistic inference, with a certain diagnosis state conditioned on k observed evidence variables X_i with the according values x_i^j:

$$P(Diag_n = OK|\, X_1 = x_1^j, ..., X_k = x_k^j) = ? \tag{4.26}$$

The inference calculation is explained exemplarily in section 3.4. Eventual temporal (un)observability of an arbitrary set of evidence variables does not influence the computability of the result. Based on the results for individual objects, the global diagnosis result can be computed for an arbitrary set of objects (e.g. for the interpreted environmental model described in section 4.2.6).

4.3.4 Environment objects

The proposed diagnosis of the perception of environment objects is based on preprocessed data of individual sensors, since this is the most common output of today's automotive sensors. The left part of the figure 4.15 presents the complete diagnosis network of the environmental perception of an object from the perspective of one sensor A. The right part of this figure outlines a combination with a second sensor B (analog structure to A) for the case of the host object being equipped with two sensors. The network structure can be extended for more sensors by adding additional child sub-network(s) to the central node "Diagnosis object n". The diagnosis perspectives of the sensors are combined under the consideration of their general plausibility. Certainly, the individual sensor sub-networks can be parameterized differently.

Furthermore, the sensors dispose of different fields of view. Therefore, if none of the data from a certain sensor can be assigned to a certain object, it must be distinguished, if this object is located in the field of view of the according sensor. Table 4.2 gives an overview of the possible combinations (cp. table 2.2). If this is not the case, the evidence variables in the according subnetwork can be left untouched (neutral). In the opposite

Sensor measurement	Field of view	Effect
not available	inside	penalizing
not available	outside	neutral
available	inside outside	depending on evidence

Table 4.2: Possible relations between sensor measurements and its field of view

case, if a sensor does not detect an object inside its field of view, it must be penalized accordingly to reflect this in the global estimate. This can be done by setting the evidence variables to the theoretical values of the tracking quality components (e.g. max. dropouts) and the association to unsuccessful.

The resulting diagnosis network has a semi-hierarchical topology. The most important factors to be modeled are the tracking data, the prerequisites of the relative state of the object and a measure for the sensor data fusion association success as discussed in section 4.3.2. In the following the meaning of the modeled variables (nodes) is summarized:

- Tracking Quality: A measure of the quality of object's following (tracking) by the according sensor.
- Dropouts: The number of tracking errors, where the object tracked by a sensor is lost and then detected again, in a defined time window.
- Tracking duration: The time period since the object was initialized.
- Association success: A comparable measure of the success of the association step, based on statistical distance.
- State prerequisites: A measure of how well the current state (estimated distance, azimuth, velocity) of the object can be measured based on previous knowledge (experience).
- Distance, azimuth, velocity: The relative state variables of the detected object determined by the according sensor.

The childless nodes correspond to the measurements or the values directly derived from the measurements. In this case, all these nodes are designated for setting of evidences (observations). The result of the diagnosis

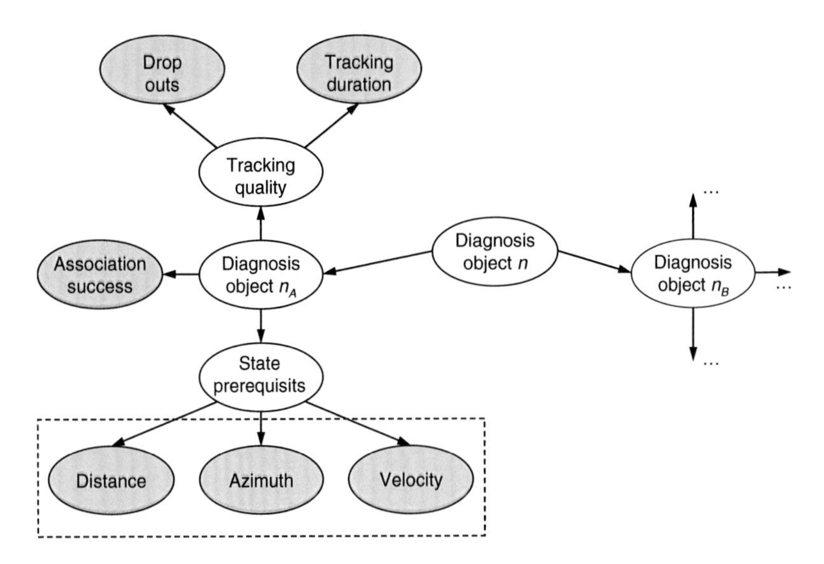

Figure 4.15: Bayesian network: diagnosis of environmental object n (use of two sensors A and B)

from the perspective of the according sensor can be determined by inference (s. section 3.4). If one or more of the (childless) variables remain unobserved, the inference can still be computed and hence the diagnosis result as well. This advantageous property is implied directly by the Bayesian networks definition (s. section 3.2).

4.3.5 Driving lane

For the purpose of diagnosis, the lane is treated as a special case of an environment object and the diagnosis of the lane detection is based on the same principle as the diagnosis of the environmental perception of environment objects. It is tracked in the analog way as the "conventional" environment objects, possibly also by more sensors. Nevertheless, there are other components within the state vector to be reasonably considered as child nodes of the "State prerequisites" node (dotted box in figure 4.16). In the case of a clothoidal lane model (s. section 4.2.3), those are particularly:

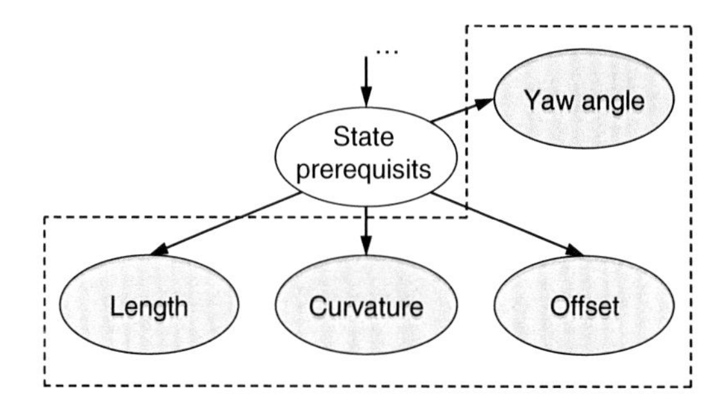

Figure 4.16: Lane subnet – clothoidal lane model

- Length: Detected length of the lane which corresponds to the farthest point used for the lane reconstruction.
- Curvature: Reciprocal curve radius, proportional to the length of the clothoid.
- Offset: Lateral displacement of the host vehicle related to the middle of the lane.
- Yaw angle: Angle between the lane and the vehicle axis.

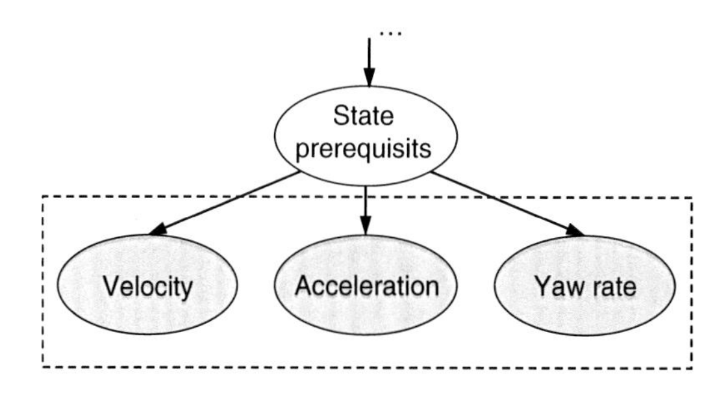

Figure 4.17: Ego subnet – host vehicle odometry sensors

4.3.6 Host vehicle

The environment objects are detected by the on-board sensors in the common case always relatively. Hence, the own movement estimation plays an important role in the environmental perception (cp. figure 4.4). The diagnosis of the host vehicle's movement estimation is based on the same principles as the method introduced for environment objects (see above). Thereby, the most of the components of the host vehicle state vector can be determined in a considerably more precise way than for the environment objects, and considered in the diagnosis. Hence, the child nodes of the "State prerequisites" node (dotted box in figure 4.17) are to be chosen differently. In particular the following components are to be reasonably considered:

- Longitudinal and lateral velocity: The distance traveled per unit of time in the according direction.
- Longitudinal and lateral acceleration: The change in velocity (in according direction) over time.
- Yaw rate: The speed of the vehicle rotation around the vertical axis.

4.4 Intent estimation

The intent estimation module is a part of the interpretation layer. It bases on the interpreted environmental model (s. section 4.2.6). By an integrated consideration of the objects of the interpreted environment model more abstract environment modeling is possible (e.g. by means of driving lanes).

Further interpretation of a traffic situation can reduce the estimated geometrical and dynamical state variables of individual environment objects to their relevant properties. Additionally, it can reduce the computational demand of the assistance applications. The desired maneuver-level based estimation assigns the environment objects a probability distribution of defined intents based on predefined behavior patterns, up-to-date knowledge and previous experience.

In this section, this method is exemplarily demonstrated by means of one of the common highway maneuvers – the lane change. Analog to section 4.3, the proposed system is represented by a Bayesian network with discrete variables. Thus, the lane change probability in the left neighbor lane of an arbitrary object can be expressed by a probabilistic inference equation as follows:

$$P(LaneChange = left|\, X_1 = x_1^j, ..., X_k = x_k^j) = ? \tag{4.27}$$

The result can be computed in accordance to the k available evidence variables X_i with the according values x_i^j as demonstrated in section 3.4.

The proposed model is based on the current state of the object of interest, the information about the available driving lane(s) and the relations between this object and other traffic participants. Thereby, a subset of the evidence variables needs to be preprocessed for the model representation.

The longitudinal relation between two objects can be described significantly by the time gap τ between both objects. Time gap represents the time difference between the passing of two consecutive vehicles a given reference point. Because the sensor range at common highway velocities is limited to few seconds, an approximation by a constant velocity model can be used for its prediction:

$$\tau = \frac{\Delta x}{v_0} \tag{4.28}$$

where Δx is the initial longitudinal distance value between both objects and v_0 the initial velocity of the following object.

The relationships between the objects in lateral direction can be intuitively described by the relative lane allocation by means of neighbor-lanes. The lane allocation of the objects is provided directly as an implicit part of the interpreted environment model (s. section 4.2.6).

The proposed Bayesian network is causally structured. Given observable characteristic features of the scenario from perspective of a defined object (e.g. time gap of the object ahead) and the observed symptoms (e.g. flashing indicator state) the Bayesian network can be used to compute the probability distribution of the hidden cause (the lane change intent).

From the host vehicle driver's point of view, this task has been examined in detail by [SG08]. The focus of this section is to demonstrate this ability in general for an arbitrary environmental object (incl. host vehicle) and its possible contribution to the presented integrated approach. Both static and dynamic Bayesian networks can be used to solve this task. This section presents a solution based on a static variant.

In the following the meaning of the modeled variables (nodes) is summarized, whereby n represents the identifier of the estimated vehicle:

- Lane change n: Lane change/keep probability of vehicle n.

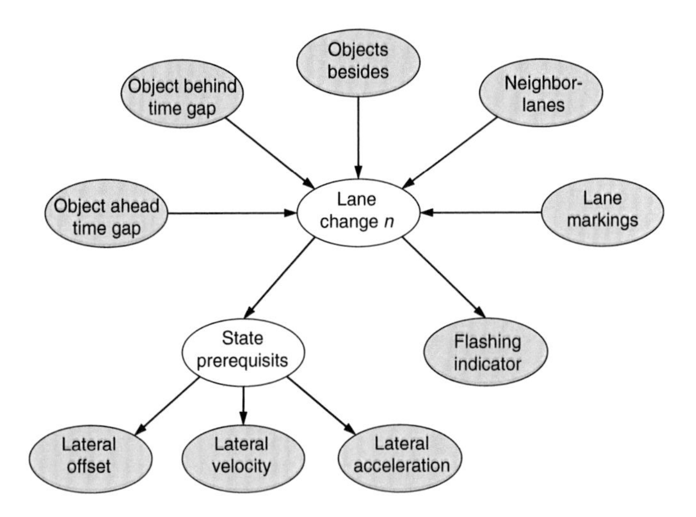

Figure 4.18: Intent estimation network

- Object ahead (behind) time gap: Time gap between the vehicle n and the first object in the according direction.
- Object besides: Presence of objects in left and/or right neighbor lane in critical area.
- Neighbor lanes: Presence of left and/or right neighbor lane.
- Lane markings: Lane markings type setup (e.g. solid, dashed).
- Flashing indicator: Status of the flashing indicator of vehicle n.
- State prerequisites: A measure of the probability of the overtaking maneuver based purely on the current relative state of the vehicle n according to the assigned driving lane (estimated lateral offset in the lane, lateral velocity, lateral acceleration) and previous knowledge (experience).

According to the general notation in this thesis the shaded nodes can be observed and the clear nodes are hidden. As mentioned above, some of the observation nodes cannot be directly "filled" with measured evidences and a preprocessing algorithm is needed (e.g. object clustering for "Object ahead" or "Object besides"). Some of today's sensors do already natively support this functionality.

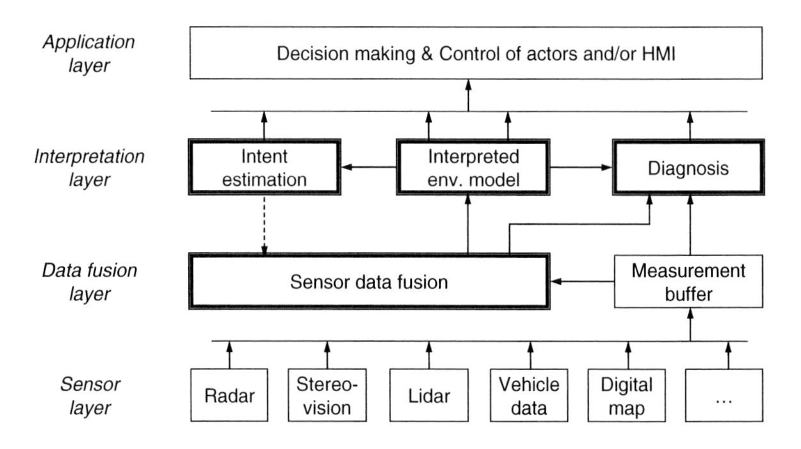

Figure 4.19: Integrated Approach functional diagram

Besides the estimated intents of detected environmental objects, the assistance applications can profit from the resulting estimates of the host vehicle driver's intents by e.g. adjusting their functionality adaptively.

Furthermore, the low level environment object's movement prediction estimated (s. section 4.2) by the tracking algorithm (according to the used dynamic model) can be improved by incorporation of additional available knowledge about the intents of the objects. For this reason, a feedback of the intent estimation module to the sensor data fusion is optionally available (dashed arrow in figure 4.19). Thus, maneuver level intent estimation can improve the prediction step.

4.5 Summary

This section summarizes the introduced probabilistic methods to get a robust environmental perception mechanism with diagnosis and intent estimation ability as proposed in section 4.1. This innovative approach covers a broad research field from sensor measurements up to advanced interpretation algorithms by means of the novel high level diagnosis and the intent estimation of traffic participants (s. table 4.3). Thereby, Bayesian networks and thus the probability serves as uniform denominator, which assures a consistent mathematical background of the entire approach.

Module	Technical innovation
Sensor data fusion	Track-to-Track data fusion approach implemented by a Bayesian network
Diagnosis	Probabilistic high level diagnosis of detected objects with respect to state variables, detection and association uncertainty
Interpreted env. model	Compact abstract description of a driving scene
Intent estimation	Full integration of a probabilistic intent estimation algorithm into the presented framework
General	Consistent extensible approach based on well-established uniform mathematical background

Table 4.3: Selected technical innovations of this chapter

Figure 4.19 reminds the block diagram of the environmental perception system proposed in the beginning of this chapter. The essential functional modules developed within the framework of this thesis are highlighted. Their working principles are described within this chapter in detail (s. sections 4.2, 4.3 and 4.4). It can be summarized as follows: The sensor data fusion module combines the sensor measurements to a virtual environment model. It incorporates an optimized track-to-track data fusion approach, which minimizes its filtering delay. The diagnosis module monitors the incoming raw sensor data and the sensor data fusion internals and estimates the environmental perception confidence level based on this information and previously acquired a priori knowledge. It considers all general sources of environmental perception uncertainty, namely the state variables uncertainty, the detection uncertainty as well as the association uncertainty. The intent estimation module performs the maneuver level intent estimation of all traffic participants based on the low level tracking information of the environmental objects and further a priori knowledge. It can be optionally coupled with the sensor data fusion module to improve its prediction step. The interpreted environment model represents a compact abstract matrix description of a driving scene with respect to the host vehicle. The output of the referred modules of the interpretation layer represents a promising information bundle, which is designated to be used and shared by future driver assistance applications.

Chapter 5

Experimental system

This chapter presents an implementation of the principles described in this thesis. Based on the objectives introduced in section 1.3, requirements for the on-board environment perception system were specified. According to this specification a system architecture was derived and implemented in an experimental vehicle (s. section 5.2). The proposed architecture was extensively practically examined (s. chapter 6).

The experimental system bases on Volkswagen Passat type B6. This vehicle was prototypically equipped with additional hardware, which implements the environmental perception system and the Integrated Lateral Assistance application.

The presented experimental system was developed within the framework of research initiative AKTIV (s. section 2.1.2), namely within its subproject "Integrated Lateral Assistance" between September 2006 and August 2010. The general objective of this subproject was a continuous lateral and longitudinal assistance in the full speed range up to 180 km/h on well-structured roads like motorways and highways as well as inside of construction sites.

The described integrated perception algorithms (s. chapter 4) require a high performance sensor set-up to be able to provide an environmental perception with high reliability. The presented sensor setup bases on a stereo camera system and radar. The sensors monitor the surrounding of the host vehicle and deliver information about the lane and other vehicles in front as well as in the side areas and in the area behind the host vehicle. Besides these environmental sensors a GNSS[1] based digital map was integrated into the system to supply the static information about the environment. Based on the sensor data a common interpreted environmental model (s. section 4.2.6) is continuously calculated, which includes

[1] Global Navigation Satellite System

Figure 5.1: Experimental vehicle

Figure 5.2: Detail of the cameras behind the windshield

the state estimation of included objects as well as their diagnosis and intent indication.

The output of the integrated perception algorithms is passed to the integrated longitudinal and lateral control instance. It processes selected objects of the interpreted environmental model and provides a joint longitudinal and lateral vehicle control. This approach leads to various advantages in comparison to the common parallel longitudinal and lateral control approaches: e.g it increases the comfort and safety in the driving situations that address longitudinal and lateral control aspects simultaneously, (e.g. curve drive situations and lane changes). Beyond the integrated environmental perception and the integrated control approach an integrated Human Machine Interface (HMI) was developed to support a joint concept for longitudinal and lateral assistance (s. figure 5.3). The aim of the consequent consideration of integrated approaches was especially to increase the transparency of the entire system and to avoid mode confusion.

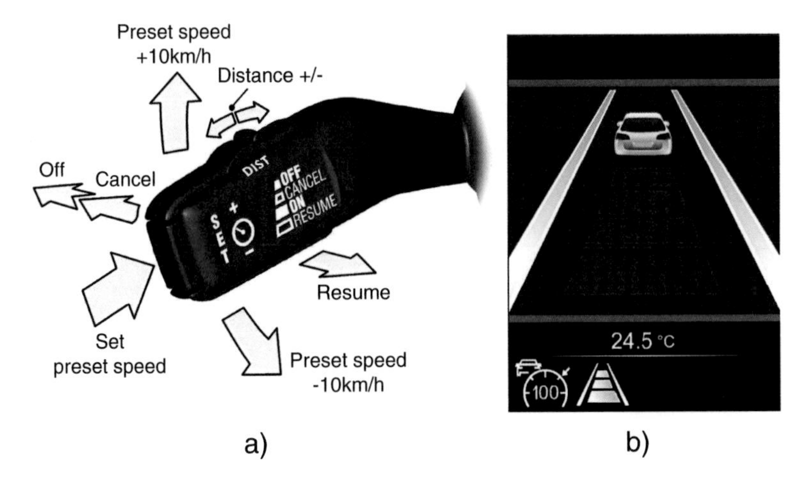

Figure 5.3: HMI of the Integrated Lateral Assistance application: a) Control lever; b) Driver information display [Eig09]

This chapter focuses generally on the environmental perception part of the system. The Integrated Lateral Assistance application and the according controlling algorithms are presented in [Eig09] in detail.

5.1 Applied Requirements

Based on section 4.2 a common perception concept was implemented. Thus all available sensor data are combined by sensor data fusion algorithms to a common environment model. This model serves as a basis for the interpretation and hence for the attached assistance applications.

The core of the sensor setup is represented by radar and by stereo camera. Due to their measurement principles, both sensors complement on each other. Radar is able to measure velocity as well as accurate longitudinal distance and it is robust against dirt and bad weather conditions. Stereo camera is able to measure lateral position and geometry of target objects precisely. Furthermore, a stereo camera system is able to detect static and/or boundary objects, that are essentially important for the desired application (s. section 1.3). These both environmental sensors build the core of the sensor system and serve as a main input for the data fusion algorithms.

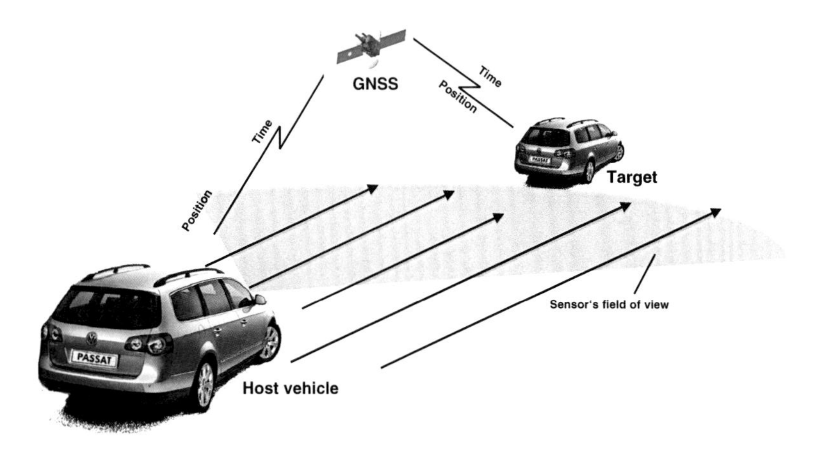

Figure 5.4: Sensor evaluation with GNSS reference

In order to achieve the required robustness, a diagnostic approach was implemented as a part of the interpretation layer, which monitors continuously the incoming sensor data and estimates the reliability of the data fusion results. Further interpretation of the traffic scenarios can be provided by means of the intent estimation of traffic participants.

Robust environmental perception based on the integrated approach presented in chapter 4 requires a detailed knowledge about the used hardware, especially sensors. In section 4.3.2 a method was introduced, which allows experimental determination characteristics of an arbitrary environment sensor. Figure 5.4 illustrates the experimental set-up. Further technical details according the reference framework are described in appendix A.

Based on the developed principle it is possible (among other parameters) to determine the real field of view of the sensors and the deviance of the sensor measurements compared to the reference platform. Figure 5.5 shows a field of view survey of one of the sensors based on automated target control according to a dense pattern of figure 4.11.

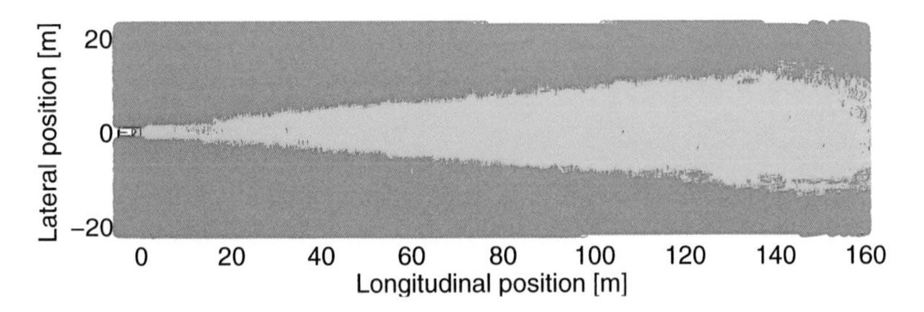

Figure 5.5: Experimentally determined field of view of a long range radar

The obtained results allow a comprehensive analysis of the sensor data and thus a substantiated statement about its performance. The obtained knowledge can be used for sensor choice as well as for the internal algorithm development.

5.2 Architecture

The developed system makes use of series hardware available in VW Passat type B6 to control the vehicle. For this purpose, the electro-mechanical power steering (EPS) and the adaptive cruise control (ACC) interfaces have been modified to be completely computer-controllable. This allows the system to control the vehicle in both lateral and longitudinal direction [Eig09, Mau00]. While the most of the car manufacturers offer similar ACC functionality in their high-end products, the electro-mechanical power steering is still a non-common notable feature. EPS is a speed-sensitive electrical power-assisted steering system which works without any hydraulic components. Compared to hydraulic power steering, vehicles fitted with electromechanical power steering benefit from lower fuel consumption and new convenience and safety functions.

The software algorithms developed within the scope of this thesis were mainly implemented in C++ within a software framework called Automotive Data and Time-Triggered Framework (ADTF) [Löb08]. This framework bases on a modular approach, where individual functions are encapsulated in software modules. The resulting structure, inspired by the Microsoft Component Object Model (COM), enables high flexibility in the development and reusage of individual software modules.

5.2.1 Hardware

The sensor system of the vehicle consisting originally of an adaptive cruise control radar 77 GHz and a lane assist mono camera was extended by additional sensors to be able to realize the desired functionality as depicted in figure 5.6. The additional sensors are fully integrated into the vehicle as shown by figure 5.1 and figure 5.2. The main scope of this sensor setup is robust detection of environmental objects in front of the vehicle, especially other traffic participants, boundary objects and the lane markings. The front sensor cluster is complemented by a pair of rear/side sensors, which detect eventual traffic approaching from the back.

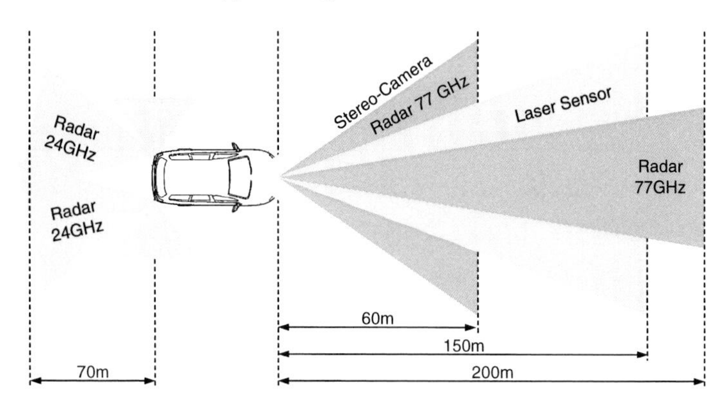

Figure 5.6: Sensor setup overview

Sensor	Field of view [°]	Range [m]	Cycle time [ms]
77 GHz front scanning radar	18 58	200 60	67
Stereo-camera system	58	60	72
Laser sensor	30	150	100
24 GHz side assist radars	2x 18 2x 120	72 blind spot	40

Table 5.1: Extended sensor setup

Table 5.1 summarizes the extended sensor setup. The 77 GHz scanning two-in-one radar covers both near and far range. The near range is additionally covered by a stereo-camera system, which is able to detect the lane markings and boundary objects (e.g. safety fences) as well. A laser sensor is used to additionally support the application in the middle range. Furthermore, the test vehicle is equipped with side assist 24 GHz radars (used in Volkswagen Phaeton, Touareg). They can detect objects approaching from the back, and cover the blind spot area up to the B-pillar of the vehicle as well. The measurements from all available sensors serve as sensor data fusion input as described in 4.2. The asynchronous architecture is illustrated by the specification of average cycle times of the sensors in table 5.1.

The data processing hardware of the system consists of multiple computers and bus gateways mounted in the trunk of the test vehicle. The required sensor data are preprocessed and forwarded to the sensor bus, which serves as a central information source for the fusion PC. A documentation camera (s. figure 5.2, in blue) connected to the fusion PC supplies video images, which can be stored together with the sensor data for documentary purpose. The output of the fusion PC is transmitted to the controller PC together with a status message from the sensor gateway, which monitors defined timing criteria. According to this information, this computer triggers the actors and the human-machine-interface of the vehicle. The PCs of the system can be accessed and controlled via remote network connection or optional peripheries. The described hardware architecture is depicted in figure 5.7. The according integration of the data processing hardware in the spare wheel recess of the trunk is presented in figure 5.8.

The stereo vision system mounted behind the windshield consists of two cameras (figure 5.2, leftmost and rightmost) and an independent PC which hosts the image processing dense stereo algorithms and an object extraction algorithm. Both cameras are triggered by a frame-grabber to produce synchronized frames. The frames are transmitted via camera link to the PC where the object state observations are extracted and outputted to the sensor bus. Thereby the output data are converted into the vehicle coordinates related to the center of the host vehicle's front bumper.

The 77 GHz combined middle and long range radar is equipped with a mechanically scanning antenna. Its principle is based on a grooved roller with a structured metal surface, which deflects the emitted electromagnetic waves in form of steerable narrow beams. Through appropriate structuring of the roller two azimuthal scans per one measurement cycle are processed:

- Long range scan, which uses a swinging beam of 2.4° via a field of 16° with 200 m range.
- Middle range scan, which swings with 8° broad beam via a field of 58° with approximatively 60 m range.

The sensor measures independently the distance and velocity (Doppler's principle) to objects in one measuring cycle. It bases on frequency modulated continuous wave (FMCW) with fast ramps. The output data is transmitted via gateway to the sensor bus.

The laser sensor scans the surrounding environment by means of an infrared laser beam that is deflected by a rotating mirror. The laser emits short pulses that are reflected by environment objects. The reflections are received by the scanner and the response times of the pulses are measured. The distances of the objects can then be determined from the response time and the known velocity of light. Moreover, the direction from which the reflected beam has been received is derived from the angular position of the rotating mirror. The reflection points are then clustered and transmitted via gateway to the sensor bus.

The 24 GHz radar system consists of two sensors mounted in the rear corners of the vehicle under the bumper at a certain mechanical squinting angle. One of the sensors acts as a master and combines the data captured by both sensors by using a private link between them. Its principle bases on 24 GHz narrow-band frequency modulated shift keying (FMSK-2) signal waveform [RM01], which provides the possibility of an unambiguous and simultaneous target range and velocity measurement. The field of

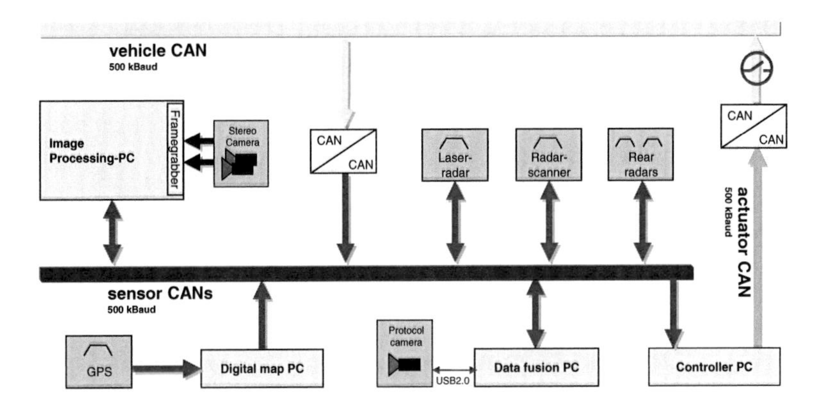

Figure 5.7: Simplified hardware architecture

Figure 5.8: Experimental vehicle: set of hardware integrated in the trunk

view covers the most of the blind spot zone, and a 70 m rear area containing three lanes in the back (left neighbor, driving and right neighbor lane). The blind spot coverage is achieved by a special antenna design in combination with detection algorithms that exploit intentional side lobes of the beam pattern. The output of the master sensor is transmitted via gateway to the sensor bus.

5.2.2 Software

The environmental data processing is done by the fusion PC mounted in the trunk using Advanced Data and Time-Triggered Framework (ADTF) version 1.1.3 by Audi Electronics Venture GmbH. Due to its modular concept with various standard components (e.g. automotive bus access tools, data recorder and offline player), ADTF provides an adequate base for the development of prototypical automotive functions and facilitates the software testing and verification process. Besides data recording, this framework provides a possibility for real-time data playback and online data processing and thus it combines a development environment with an interactive work environment.

The individual modules in the ADTF nomenclature are called "filters". A filter is a software component implemented in the programming language C++ with precisely defined input and output interfaces, referred to as "pins". The resulting algorithm consists of filters connected via compatible pins, that build a so-called "filter-graph". The ADTF infrastructure arranges automatically the output data of a source filter to reach the according connected destination filter. The exchange of data between the various pins of the filters is performed by transmitting and receiving of abstract data packets called "media-samples". The data flow is generally processed serially: A transmit method of the source module initiates directly the receive method of the target module.

Figure 5.9 shows a simplified filter-graph of the proposed system (cp. figure 4.1, figure 4.3). The bus device is responsible for the external interface (e.g. sensor bus). The data decoder interprets the according protocols and maps the included data on defined sensor data structures. The data are transmitted via pins to the according data fusion algorithms. The ego data fusion supplies the host vehicle motion data for the remaining fusion modules. The interpretation subgraph in figure 5.9 is composed of modules, that process the tasks described in section 4.2.6, section 4.3 and section 4.4 and generate a consistent data output.

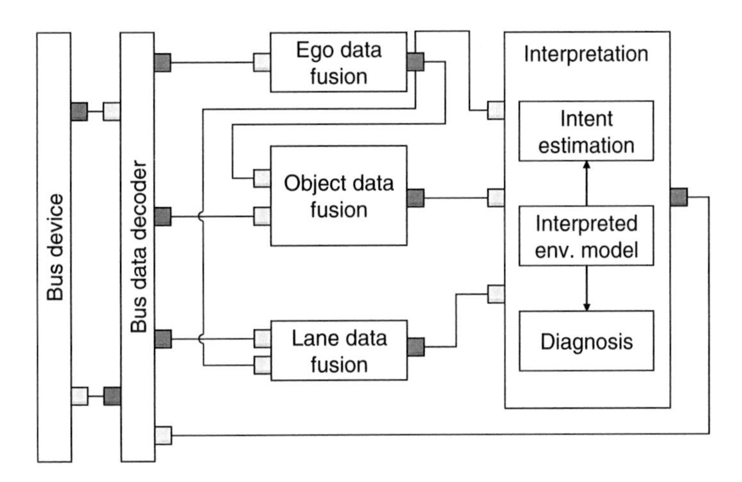

Figure 5.9: Simplified filter-graph; little dark gray boxes represent output pins, little light gray boxes represent input pins

The described functionality suites for the online mode, where the system processes the data from on-board systems and transmits the results to the controller PC. As mentioned above, by adding of a "hard disk recorder" filter to the system, it is possible to capture data for later offline analysis. In the offline mode, the bus device is substituted by a "hard disk player", which replays the data as it was captured by the hard disk recorder. This is an important feature, which allows efficient offline algorithm development.

The implementation of Bayesian network algorithms was supported by a structural modeling, inference, and learning engine (SMILE) version 2.0 developed by the University of Pittsburgh. SMILE is a comprehensive library of C++ classes implementing graphical probabilistic and decision-theoretic models. It is represented by a dynamic link library (DLL). The used discrete probabilistic models (s. sections 4.1 and 4.4) were developed in its graphical interface and SMILE is utilized as back-end inference engine called from ADTF.

The extensive analysis of sensor characteristics was performed offline using Matlab version 2008b by The Mathworks. It provides a high-level language and interactive environment with sophisticated mathematical engine and advanced visualizing possibilities. Within Matlab, a software tool was implemented, which largely automates the data processing and the analysis of sensor characteristics (s. section 4.3.2). It is able to output a variety of visualized results, which allows the parameterizing of the system models and further examination. The proposed concept is intended for use with potential further sensors.

Chapter 6

Evaluation and Results

This chapter investigates the performance capability and simultaneously the limits of the experimental system described in chapter 5. Thereby the performance criteria are derived from the objectives specified in section 1.3. The integral aim of the presented approach is a precise estimation of the environment scenario together with continuous diagnosis of all general sources of uncertainty by means of a confidence level assigned to each detected environment object. Based on this information, the assistance application(s) can support the vehicle control safely and react to potential issues by e.g. appropriate functionality adaptation or information output to the driver via human machine interface.

Due to sensor hardware specifics and tight coupling of all functional modules a potential simulation-based evaluation was excluded. In this case, simulation cannot completely replace practical measurements in form of driving tests. The presented results are based on real test drives in the local proving ground, which allows safe testing of arbitrary driving maneuvers.

6.1 Test scenarios

Within the scope of "Integrated Lateral Assistance" project, a test catalog was created [GEJS08], which contains the applied requirements by means of defined function relevant test scenarios with focus on parallel traffic. The most of the specified scenarios can be separated in particular relative lateral and longitudinal maneuvers with according parameters.

The evaluation of the environmental perception concept focuses on the worst case analysis examining the area close to the system limits. Based on this assumption test scenarios have been selected for the presentation through this chapter. They contain rapid movements of a target vehicle relative to the host vehicle in both longitudinal and lateral direction by means of intensive brake application and a rapid lane change maneuver. The results can be generalized for the entire system specification definition. In the following the experimental procedure is described in detail.

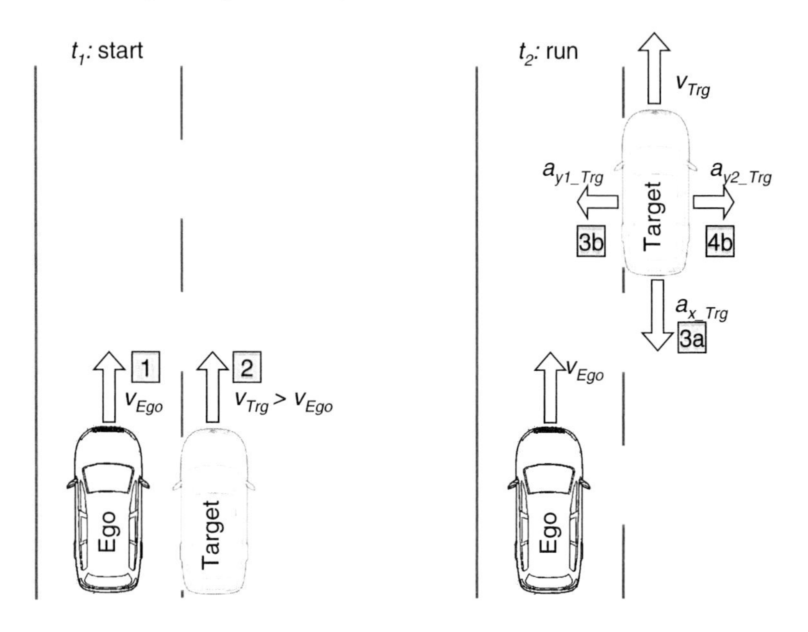

Figure 6.1: Visualization of test scenarios: option a) Target vehicle varying its velocity with constant lane offset; option b) Target vehicle varying its lateral offset by means of a lane change maneuver

At the beginning of both test scenarios the host (ego) and the target vehicle are placed stationary next to each other. The host vehicle is placed central in a straight driving lane and the target vehicle is situated in the left part of the right neighbor lane. At time t_1, the host vehicle accelerates smoothly to reach the desired velocity v_{Ego}. Then, the target vehicle accelerates and follows the host vehicle with constant lateral offset and a significantly higher goal velocity v_{Trg} (s. figure 6.1, left).

After certain time, the target vehicle passes the host vehicle from the right and reduces its velocity smoothly. At time t_2, the target vehicle is located in front of the host vehicle, with neutral relative velocity ($v_{Trg} = v_{Ego}$), still with a constant lateral offset (s. figure 6.1, right). At this time, the target vehicle operates in accordance to one of the following options depending on the intended scenario:

Scenario a) Target vehicle reduces further its longitudinal velocity v_{Trg} by a deceleration a_{x_Trg} caused by intensive brake application. As a result its relative velocity becomes negative ($v_{Trg} < v_{Ego}$) and the host vehicle passes the target vehicle in turn. In this scenario, the lateral offset of the target vehicle remains constant relative to the host vehicle. By use of the symbols from figure 6.1, the entire scenario procedure can be summarized with the following sequence: $\boxed{1} \rightarrow \boxed{2} \rightarrow \boxed{3a}$.

Scenario b) Target vehicle eliminates its lateral offset by a lateral acceleration a_{y1_Trg} as a result of a rapid lane change maneuver. Subsequently, it performs the inverse maneuver with a_{y2_Trg} and changes back to its initial lateral position. The relative longitudinal velocity (and distance) remains constant. By use of the symbols from figure 6.1, the entire scenario procedure can be summarized as follows: $\boxed{1} \rightarrow \boxed{2} \rightarrow \boxed{3b} \rightarrow \boxed{4b}$.

The proposed scenarios cover the main aspects, which are to be examined according to the desired application. They include parallel traffic with variable lateral offset, continuous (positive, neutral and negative) lateral and longitudinal velocity difference as well as significantly accelerated and decelerated movements with respect to the desired application and the according system specifications. Furthermore the target object comes in and leaves the field of view within the tests.

6.2 Sensor data fusion

The presented multi-sensorial environmental perception approach bases on sensor data fusion algorithms. It performs an event controlled processing of the sensor data. As a result, a virtual environment model is constructed. In this model the entire information about the host vehicle and its environment is accumulated.

Figure 6.2 depicts the dependencies between the individual data fusion modules (s. section 4.2). The ego data fusion module provides source data for the lane and object data fusion algorithms. The results from all data fusion modules are further processed by the interpretation layer. In the following, the data fusion modules are examined individually based on the test scenarios introduced in section 6.1.

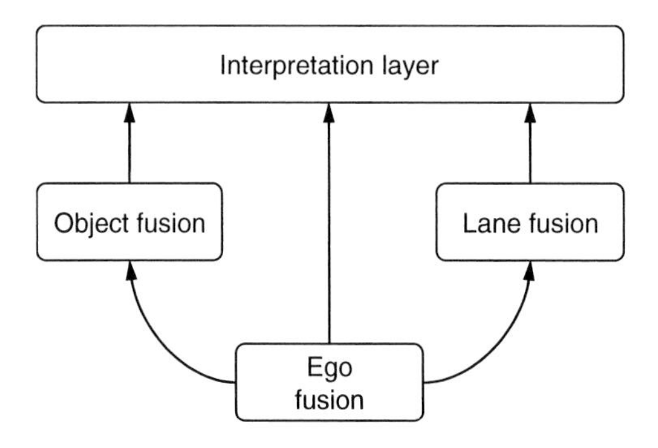

Figure 6.2: Data fusion modules dependency scheme

6.2.1 Object data fusion

This section presents an analysis of classical (pseudo-)measurement data fusion approach and the track data fusion approach derived in section 4.2.5. Both algorithms are used to combine identical pre-tracked sensor data from a stereo camera system, laser sensor and scanning radar (s. chapter 5).

During the test drives, both the host and the target vehicle were equipped with a reference system (s. appendix A), which allows an objective evaluation. In the following selected representative results are presented and discussed exemplarily in detail. Moreover, an accumulative statistics over the total set of test drives is included, which focuses on the estimation error of both approaches, their smoothness, and latency.

Figure 6.3 (top) shows the relative longitudinal distance of the target vehicle over time by means of the first scenario a), where it variates its longitudinal velocity by an intensive brake application. Besides the source sensor data the output of both fusion algorithms is shown as well as the reference data. The area where the longitudinal velocity ratio of both vehicles rapidly changes is magnified. The bottom part of the figure depicts the availability of individual sensor data. One can notice the track level and the measurement level fusion algorithms are in accordance with the reference and do not differ significantly due to the generally accurate (longitudinal) distance measuring ability of the used sensors (esp. laser and radar). Nevertheless, in high dynamic scenarios (i.e. object initialization, rapid state change) one can recognize minor differences.

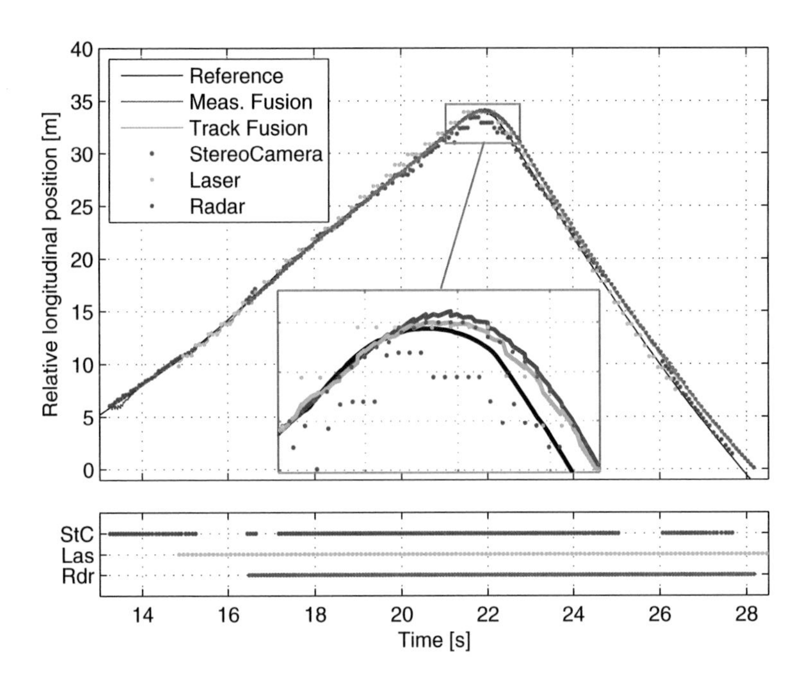

Figure 6.3: Test scenario a): relative longitudinal position over time from sensors and fusion algorithms with magnified inflection area (top); and the sensor availability (bottom)

Figure 6.4 (top) shows the initial phase of the same scenario as figure 6.3 in detail. While the track level fusion result continuously dynamically oscillates around the correct solution the measurement level fusion needs a defined amount of time to converge to the right solution with low frequency and overshooting tendency. In this example, the initial deviance of the measurement fusion is mainly caused by a wrong velocity guess at object initialization, which assumes no velocity (stationary) due to unavailability of velocity measurements from the stereo camera system. The magnified area focuses on the influence of radar measurements availability on the

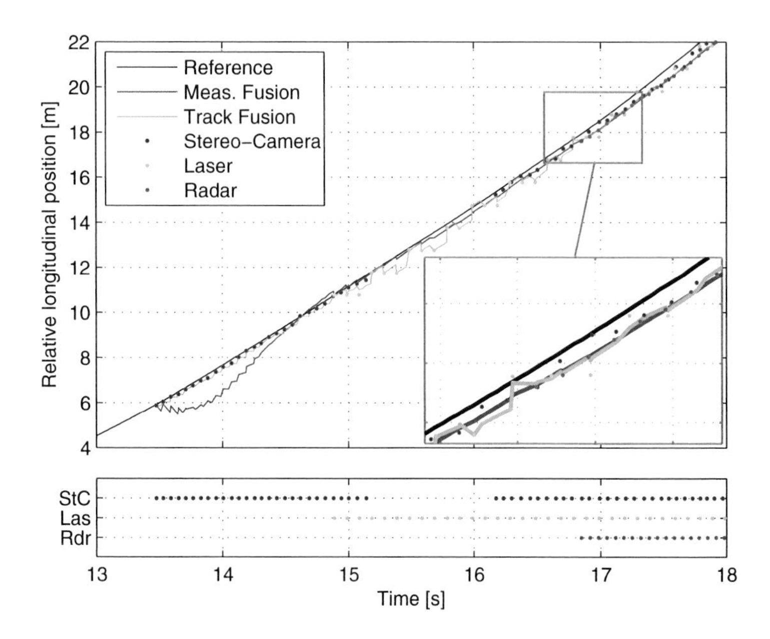

Figure 6.4: Test scenario a): initialization phase detail: Relative longitudinal position over time from sensors and fusion algorithms with magnified radar initialization area (top); and the sensor availability (bottom)

fusion algorithms. It provides a stabilizing effect due to its direct velocity measurement ability, especially for the track level fusion.

In figure 6.5, relative lateral distance of the target vehicle over time by means of the second test scenario b) is presented. In this case, the target vehicle varies its lateral offset by two rapid lane change maneuvers in sequence. First, it changes from the right neighbor lane into the host vehicle lane and subsequently it changes back to the original lane. The data sources and the fusion algorithms are in accordance with the previous scenario a). The area where the relative lateral velocity direction of the target vehicle rapidly changes is magnified. One can notice, the latency of the sensor data becomes more evident than in the longitudinal scenario. The sensors are obviously designed mainly for longitudinal application (e.g.

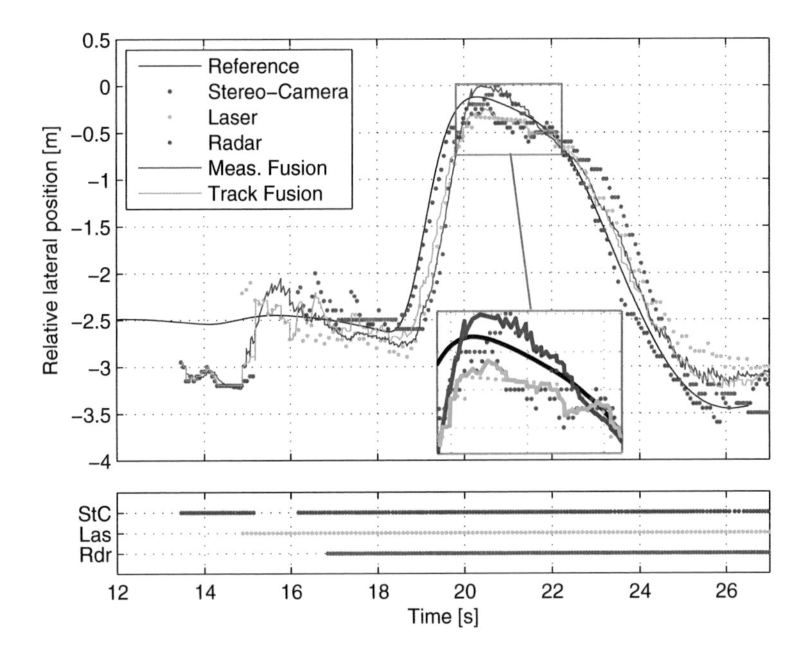

Figure 6.5: Test scenario b): relative lateral position over time from sensors and fusion algorithms with magnified inflection area (top); and the sensor availability (bottom)

ACC). As a result, on the one hand, the latency of the track level fusion is lower than the measurement fusion latency. On the other hand, the output of the track level fusion contains changes of higher frequency with a limited differentiability. The overshooting tendency of the measurement fusion from previous scenario can be confirmed.

Table 6.1 presents an accumulative statistics of all performed test drives according to scenario a). The "estimation error" section presents the average maximum, mean, and standard deviation of the particular (lateral or longitudinal) position estimation error of both algorithms according to the reference measurements. For a particular test drive and a defined fusion algorithm, the estimation X^{error} error can be calculated as follows:

$$X^{error} = \frac{1}{n}\sum_{i=1}^{n} abs(X_i^{fusion} - X_i^{ref}) \tag{6.1}$$

Thereby, the test drive is assumed to contain n fusion steps and the reference measurements X^{ref} are synchronized in accordance to the data fusion output X^{fusion}.

The "spread factor" section focuses on the smoothness of the algorithm results by means of the maximum, mean and the standard deviation of the difference between the original output of an algorithm X^{fusion} and its zero-phase smoothed output X_i^{smooth} processed in both directions. For a particular test drive and data fusion algorithm, this is defined as follows:

$$X^{spread} = \frac{1}{n}\sum_{i=1}^{n} abs(X_i^{fusion} - X_i^{smooth}) \tag{6.2}$$

Finally, the "latency" section expresses the maximum and the mean average delay of the algorithms according to the reference. For a particular test drive of and a defined fusion algorithm the following relation applies:

$$\forall t_i : X^{fusion}(t_i + t_{lat}) \cong X^{ref}(t_i) \tag{6.3}$$

This can be interpreted as a size of a temporal displacement. For each test drive a best-suited t_{lat} can be estimated.

	Estimation error [m]			Spread factor [m]			Latency [ms]	
	max	mean	std	max	mean	std	max	mean
Track fusion	1.03	0.29	0.20	0.43	0.08	0.07	159	128
Meas. fusion	0.92	0.28	0.21	0.27	0.04	0.03	163	131

Table 6.1: Scenario a): accumulative statistics of evaluation results, relative longitudinal position analysis

The values of table 6.1 can be interpreted as follows: The accuracy of the proposed track fusion algorithm is comparable to the accuracy of the classical measurement fusion algorithm. The measurement fusion output is smoother due to generally lower values of the spread factor and their lower standard deviation. Nevertheless, on the other hand, the latency values of the measurement fusion are generally higher.

The contents of table 6.2 are in accordance with table 6.1 (see above). It presents an accumulative statistics according to scenario b). Apart from the average maximum values, the average estimation error of track fusion is about 10% lower than measurement fusion with even lower standard deviation. The measurement fusion output is smoother due to a slightly lower average spread factor. Nevertheless the track fusion reaches lower latency by about 7%.

	Estimation error [m]			Spread factor [m]			Latency [ms]	
	max	mean	std	max	mean	std	max	mean
Track fusion	0.74	0.23	0.19	0.34	0.05	0.05	218	214
Meas. fusion	1.06	0.26	0.24	0.20	0.03	0.03	232	230

Table 6.2: Scenario b): accumulative statistics of evaluation results, relative lateral position analysis

As shown in this section, both data fusion approaches have their strengths. The choice of the suitable algorithm depends on a lot of factors, especially on the desired application and the characteristics of the used sensors. Basically, it will always be a trade-off between the output smoothness and the latency.

6.2.2 Ego data fusion

Supplying data for all other data fusion modules (s. figure 6.2), the ego data fusion is of fundamental importance. Eventual movement estimation error of the host vehicle causes a faulty prediction of the environmental object state and – according to the sensor measurement variance and the process noise – negatively influence the estimated state of the object as depicted in figure 4.4.

The evaluation of the on-board odometry measurements (esp. longitudinal velocity and yaw rate) approved the used sensors to be very accurate. Nevertheless, the integration of the output values can theoretically still cause a

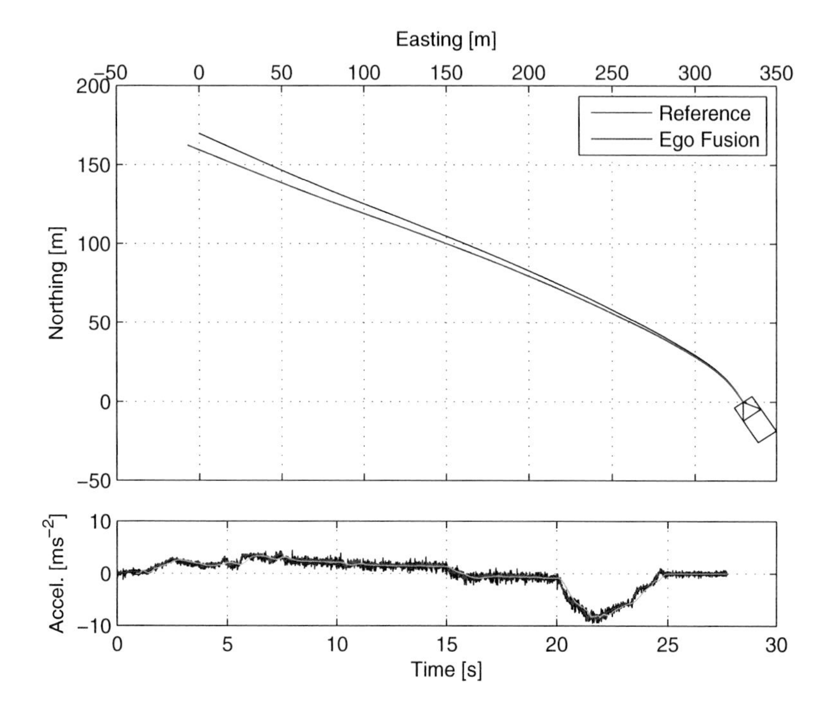

Figure 6.6: 2D ego fusion position estimation drift (top); and the according acceleration course over time (bottom)

potential degradation of the fusion results. Therefore series of test drives have been performed. Figure 6.6 shows a representative example with variable acceleration including a full brake application (s. figure 6.6, bottom). The scenario curvature is limited according to the specification of the Integrated Lateral Assistance application. Top of the figure 6.6 depicts the ego fusion output and reference data in absolute coordinates.

This test drive took 27.73 s. The final position deviation of the host vehicle was 10.16 m, thereof 7.74 m lateral (E_{lat}) and 2.42 m longitudinal (E_{lon}) contribution. The lateral contribution is related to the observed orientation deviance. The final orientation error E_{ang} was 1.3°.

$$\overline{E}_{total} = \overline{E}_{trans} + \overline{E}_{rot} = \frac{\sqrt{(E_{lat} + E_{lon})} + \tan(E_{ang}) \cdot S_{max}}{\Delta t} \tag{6.4}$$

The above absolute values imply an average translational position error between two sensor measurements (with max. 100 ms cycle time) of 2.79 cm in lateral and longitudinal 0.87 cm. The orientation error results in an additional rotational error, which causes an average lateral displacement of 1.64 cm in the maximal sensor detection range S_{max} of 200 m.

	Norm. position error [m/s]		
	max	mean	std
Ego fusion	0.456	0.189	0.113

Table 6.3: Accumulative statistics of evaluation results, position estimation error per second

The actual host vehicle position estimation deviance depends on a number of factors. The above calculation presents rather pessimistic error values between two object data fusion cycles. Table 5.1 presents the accumulative statistics of evaluation results according to the position estimation error per second (ten maximal sensor measurement cycles). The total displacement error possibly caused by inaccurate ego position estimation is within few centimeters per cycle and therefore comparable or less significant than the measurement noise of the used environment sensors.

6.2.3 Lane data fusion

The camera system of the experimental vehicle is able to track solely one complete driving lane (consisting of two lane markings) simultaneously. The test scenario used in this section represents an extreme case of scenario a) (s. section 6.1), where the target vehicle offset and width disallow the host vehicle to pass comfortably without adapting its offset in the lane. This scenario demonstrates a typical use-case of Integrated Lateral Assistance.

While the host vehicle performs the lateral maneuver inside the own lane to adapt its lateral offset, the unavailability of camera based lane measurements is pretended. In the meanwhile, the camera data is used for reference. Figure 6.7 depicts the lateral offset of the host vehicle relative to the lane. In the area between two dashed green lines, the lane fusion generates a synthetic lane based on previous measurements and the ego motion.

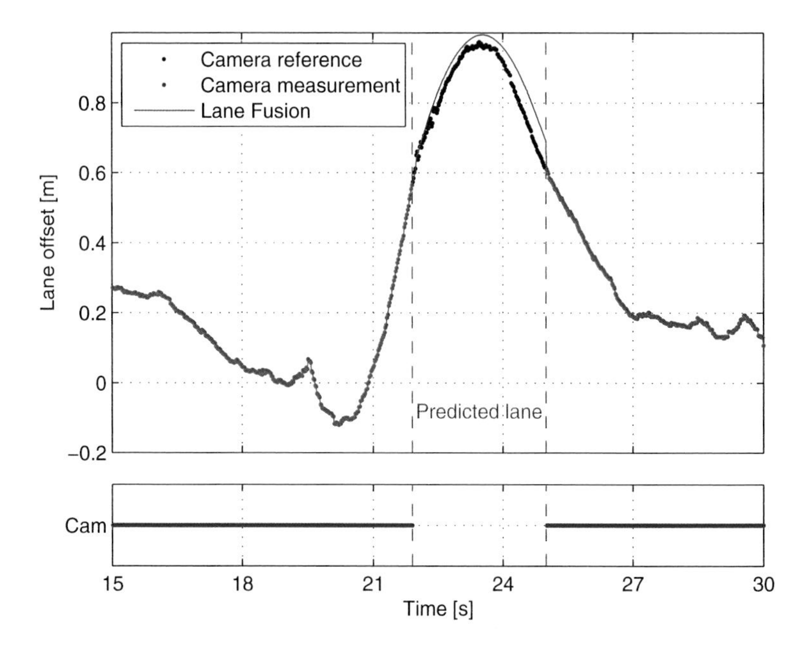

Figure 6.7: Estimated lateral offset to the lane center over time (top); and the sensor availability (bottom)

One can notice the estimated offset of the host vehicle corresponds to the according reference camera measurements. The slight discontinuity of the lane fusion output at the end of the predicted area is caused by lane model assumptions and ego position inaccuracy (s. section 6.2.2).

Table 6.4 depicts the accumulative statistics of evaluation results by means of lane offset estimation error according to the reference camera measurement. It depicts the average values over the set of relevant test drives. Because of the various "bridging" times, the values were normalized per one second interval according to the following equation:

$$E_{total} = \frac{O_{fusion} - O_{ref}}{\Delta t} \tag{6.5}$$

Thereby, the absolute error E_{total} can be computed as a difference of the fusion output O_{fusion} and the reference offset O_{ref}, normalized over the

	Norm. offset error [$m\,s^{-1}$]		
	max	mean	std
Lane fusion	0.24	0.15	0.06

Table 6.4: Accumulative statistics of evaluation results, normalized lane offset estimation error

prediction time Δt.

Generally, minor inconsistencies in the order of few decimeters during the maneuver finalization can be smoothed successfully by the vehicle controller [Eig09].

Hence, short temporal unavailability of lane measurements of the order of few seconds can be generally bridged successfully by the proposed lane fusion algorithm. In practice, such situations are not rare. They are often caused by maneuvers with relatively high offset where the host vehicle comes close to (or beyond) the lane bounds (e.g. avoidance maneuver, lane change) or temporal occlusions.

6.3 Self-diagnosis

The diagnosis module estimates the perception quality of detected environment objects. The computation bases mainly on raw sensor data, internals of the data fusion algorithm and previous experience (s. section 4.3). As a result, each object is assigned a probabilistic confidence level, a comprehensive measure which incorporates the following domains of uncertainty:

- State variables uncertainty
- Detection uncertainty
- Association uncertainty

The basic idea underlaying the proposed diagnosis approach is to respect the fact the sensors cannot detect all possible targets under all possible circumstances (e.g. misalignment, bad weather conditions, etc.). The objective is to recognize such situations and to initiate an appropriate system reaction if needed.

6.3.1 Evaluation method

The proposed evaluation method can be summarized as follows: Initially, possible (environmental) influence factors of sensor perception quality are identified. Then, selected influence factors are (synthetically) provoked during the test drives according to scenarios from section 6.1. Thereby the resulting confidence level is computed by the proposed diagnosis module based on Bayesian networks (s. section 4.3). In parallel, a classical existence estimation algorithm presented in section 2.3.2 is executed. Finally, the results of both algorithms are compared and confronted with the estimated spatial standard deviation (s. section 2.3.2) computed by a measurement fusion algorithm and the spatial error of the data fusion result according to the reference. The evaluation is performed by means of a detailed analysis of selected representative test runs and a statistical summary.

The following items represent possible common causes of potential environmental perception system issues related to individual sensors:

- Partly degraded sensor view
- Fully degraded sensor view (blackout)
- Decalibrated sensor's alignment

In case of a degraded sensor view the performance of a sensor is limited in a certain area of its field of view. The extreme case with strong limitations in the complete field of view – when the sensor cannot detect any objects at all – can be denoted as sensor "blackout". An eventual sensor decalibration can be of variable strength as well. Extreme calibration error could even cause additional ghost targets due to unsuccessful association.

The "visibility" of a particular target object depends on many factors, which are strongly influenced by the particular sensor measuring principle. Optical sensors are e.g. significantly more sensitive to measures which result in reduced sight distance (e.g. rain, dirt) than radar based sensors. Furthermore, the object detection quality depends on particular target properties. Selected potential causes of a degraded sensor view together with object features, which can influence its "visibility" itself, are listed in the table 6.5.

In the following scenarios the laser sensor is completely disabled (to make the system less robust, resp. more sensitive to outside influences) and hence the core of the sensory system, consisting of a radar and a stereo

Sensor	Degraded view	Reduced target "visibility"
Stereo camera	Rain, fogged windshield	Periodic (horizontal) or homogeneous pattern
Radar	Heavy dirt, ice	Reduced reflectivity (small cross section)
Laser	Rain, dirt, fog	Surface of reduced reflectivity

Table 6.5: Selected actions of particular sensor perception degradation, concerning the sensors view and the target object properties

camera system, is tested. Thereby, the experimental system is exposed to the identified potential influence factors of sensory perception (s. table 6.5) and the confidence level of the self diagnosis module is examined and confronted with the results of other techniques as described in section 2.3.2. The chosen test scenario bases on scenario a) from section 6.1.

6.3.2 Optimal environment conditions

Initially, the experimental system is tested according to scenario a) from section 6.1 under "optimal" conditions. Thereby the weather is cloudy, no rain, optimal sensor view and a passenger vehicle represents the target (s. figure 6.8). Therefore the sensory system is generally expected to perform well. The result of this test can be treated as reference relating to other scenarios with provoked potentially degrading factors of sensor perception.

In the following, the essential moments of this test run are described in detail (s. figure 6.9):

a) At time t=12.8 s the target object is initially detected by the stereo-camera system. Shortly after, at time t=14.5 s, the target object is initially detected by radar. Then it increases its distance due to its higher relative velocity.

b) At time t=16.2 s the target object reaches the next higher relative distance level. Shortly after it starts to decelerate.

Figure 6.8: Scenario a): optimal environment conditions; perspective of the reference camera

c) At time t=22.2 s the target object reaches the next lower relative velocity level. Its distance is therefore decreasing.

d) At time t=26.9 s the target object reaches the next lower distance level and its distance is further decreasing.

e) At time t=31.4 s the target object finally leaves the detection range of the sensors.

Figure 6.9 consists of three sub-figures. The upper sub-figure shows the confidence level assigned to the target object by both tested algorithms over time (based on the first scenario, where it variates its longitudinal velocity through initial acceleration and subsequent brake application). The middle sub-figure depicts the measurement fusion position error according to the reference and the appropriate standard deviation resulting from the fusion algorithm over time. The bottom sub-figure gives an overview about the availability of both sensors detections – stereo camera and radar abbreviated as 'StC' and 'Rdr' – over the time.

One can notice both algorithms estimating the confidence level to be qualitatively in accordance. Nevertheless, the existence estimator (s. section 2.3.2) converges quickly after the target object initialization to its max-

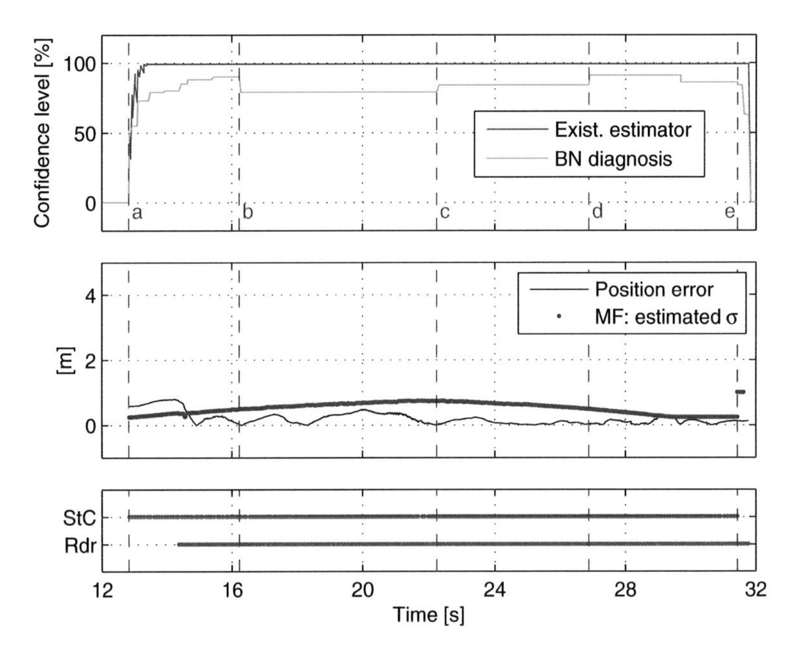

Figure 6.9: Scenario a): optimal environment conditions; confidence level estimation (top); spatial standard deviation by measurement fusion algorithm and the estimation error according to the reference (middle); sensor availability (bottom)

imal value and holds it for the entire measurement, while the diagnosis algorithm based on Bayesian networks (s. section 4.3) output offers an additional resolution in this area. The spatial sigma characteristics (s. section 2.3.2) of the measurement fusion output is continuous with moderate values as well as the fusion error according to the reference. The sensors deliver measurements continuously within their fields of view.

6.3.3 Difficult environment conditions

In this section, selected exemplary tests according to scenario a) from section 6.1 under "difficult" conditions are presented. Thereby more selected perception degrading factors are combined simultaneously (s. figure 6.10):

On the one hand, a significant fog film obscures the windshield in front of the used cameras. On the other hand, a passenger vehicle with its rear part covered by 77GHz absorbing mats was used as target. Thus the view of the camera system behind the windshield is significantly degraded as well as the reflectivity of the target vehicle from the point of view of the radar. Intuitively, this scenario setup is expected to be more challenging for the sensory system than the one described in section 6.3.2.

Figure 6.10: Scenario a): difficult environment conditions; camera perspective

In the following, a detailed analysis of the algorithm is presented by means of the first example test run (s. figure 6.11):

a) At time t=16.8 s the target vehicle is initially detected by radar. The target increases its distance due to its higher relative velocity.

b) At time t=17.5 s the target vehicle is initially detected by the stereo-camera system and its distance is still increasing.

c) At time t=19.0 s the confidence level decreases as a consequence of reaching the next higher distance level by the target object and subsequent initial dropouts of the stereo-camera system.

d) At time t=20.4 s the stereo-camera dropout rate reaches its next higher (maximum) level.

e) At time t=21.4 s the confidence level increases as a consequence of reaching the next lower relative velocity level and sparse detections by the stereo camera system. The distance of the target object is now decreasing.

f) At time t=22.0 s the target object reaches the next lower relative distance level.

g) At time t=24.2 s the confidence level decreases as a consequence of increased stereo camera system dropout rate and the next higher relative angle reached by the target object.

h) At time t=25.6 s the target object finally leaves the detection range of the sensors.

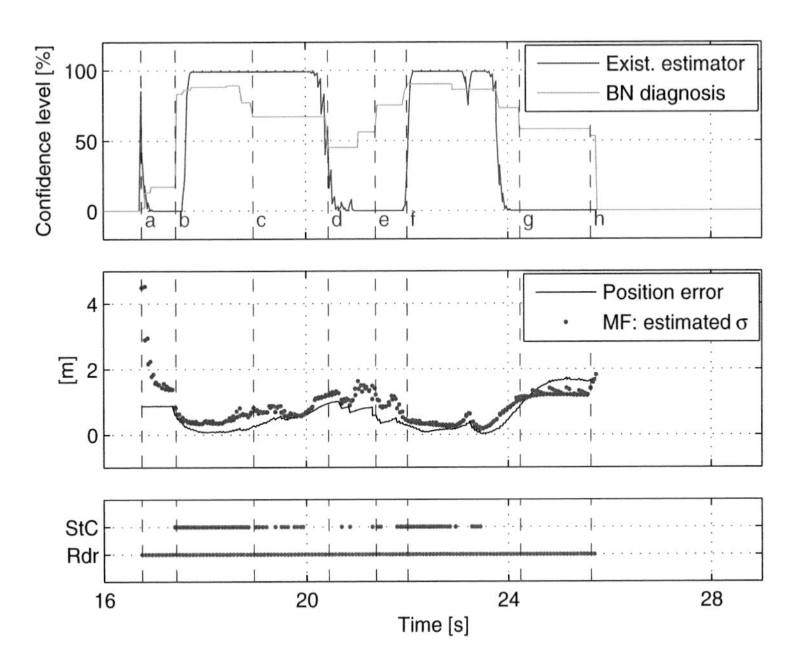

Figure 6.11: Scenario a) I: difficult environment conditions, disturbed view, target masked with absorbing mat; confidence level estimation (top); spatial standard deviation by measurement fusion algorithm and the estimation error according to the reference (middle); sensor availability (bottom)

By analogy to figure 6.9, the figure 6.11 consists of three sub-figures. The upper sub-figure depicts the confidence level assigned to the target object by both tested algorithms over time. The middle sub-figure shows the measurement fusion position error according to the reference and the appropriate standard deviation resulting from the data fusion algorithm over time. The bottom sub-figure presents the availability of both sensors detections – stereo camera and radar abbreviated as 'StC' and 'Rdr' – over the time.

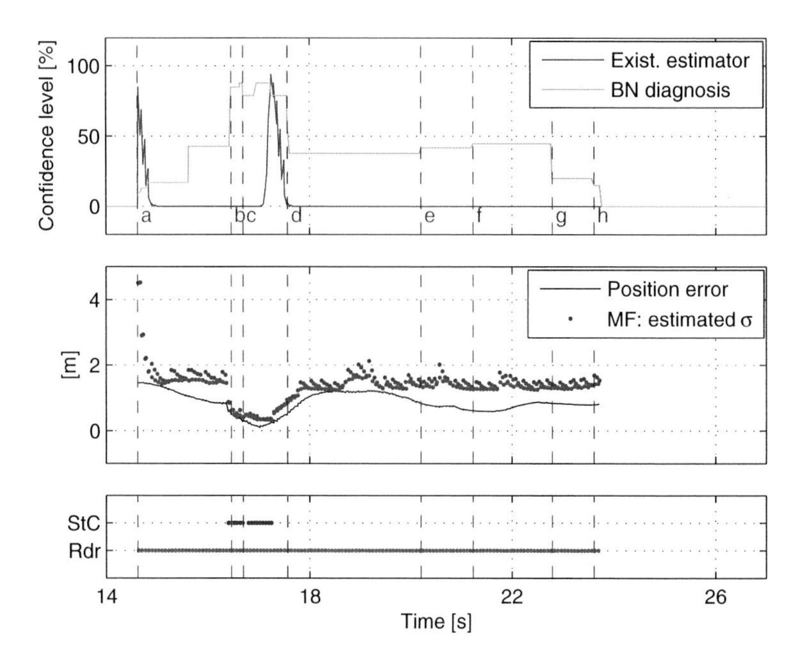

Figure 6.12: Scenario a) II: difficult environment conditions, disturbed view, target masked with absorbing mat; confidence level estimation (top); spatial standard deviation by measurement fusion algorithm and the estimation error according to the reference (middle); sensor availability (bottom)

One can notice both algorithms estimating the confidence level to be qualitatively roughly in accordance. Nevertheless, the existence estimator (s. section 2.3.2) converges quickly after the target object initialization to its maximal value and holds it for the entire measurement, while the diagnosis algorithm based on Bayesian networks (s. section 4.3) offers an additional resolution in this area.

Figure 6.12 illustrates the same scenario under similar conditions as depicted in figure 6.11 to demonstrate the spread of the results. By analogy to figure 6.11, the figure 6.12 consists of three subfigures. The upper subfigure depicts the confidence level assigned to the target object by both tested algorithms over time. The middle subfigure shows the measurement fusion position estimation error according to the reference and the appropriate standard deviation resulting from the data fusion algorithm over time. The bottom subfigure presents the availability of both sensors' detections – stereo camera and radar abbreviated as 'StC' and 'Rdr' – over the time.

Compared to figure 6.11, the performance of the stereo camera system in figure 6.12 is extremely limited to a relatively short interval in the target departing phase of the scenario. This effect causes the rising edge of the BN diagnosis approach (marking 'b') to be delayed. In the meanwhile, the existence estimator gets saturated due to the missing stereo camera measurements. Therefore the difference between the rising edges of both algorithms is increased compared to figure 6.11. In contrast, the falling edge (marking 'd') of the existence estimator arises in this case relatively earlier because this algorithm could not saturate in this short interval. In the remaining test run, a single radar sensor is not able to exceed the fifty-fifty confidence level of the diagnosis, and the existence estimator does not leave the zero value.

In order to statistically evaluate the correctness of the results of both algorithms, the following assumption was made: A positive result should reach at least the "expected" level of confidence (75%, s. figure 3.5). The evaluation bases on the relative position accuracy, which can be clearly determined by the reference measurement system. Thereby, the maximal acceptable position error of 0.5 m is defined. Thus, this definition results in the following cases and applied criteria:

- True positive: Confidence level at least at "expected" (75%), reference error lower than 0.5 m.
- True negative: Confidence level lower than "expected" (75%), reference error higher than 0.5 m.
- False positive: Confidence level at least at "expected" (75%), reference error higher than 0.5 m.
- False negative: Confidence level at lower than "expected" (75%), reference error lower than 0.5 m.

	True positive	True negative	False positive	False negative
BN diagnosis	96%	97%	4%	3%
Exist. estimator	91%	96%	9%	4%
MF σ-estimation	95%	77%	5%	23%

Table 6.6: Accumulative statistics of evaluation results, self-diagnosis

Table 6.6 presents an accumulative statistics of evaluation results of the proposed diagnosis method based on Bayesian networks, the classical existence estimator and the standard deviation estimation of a measurement fusion algorithm according to the above definition. It bases on the temporal ratio of the according cases within the performed set of test drives. It confirms the theoretical properties of the existence estimator algorithm as well as of the diagnosis approach based on Bayesian networks. The existence estimator's performance is limited to the detection uncertainty. The measurement fusion standard deviation estimation is limited to the state variables uncertainty. The proposed diagnosis involves both and the association uncertainty on the top. This fact explains the listed results, that are very promising.

6.4 Intent estimation

The intent estimation module bases on the interpreted environmental model (s. section 4.2.6). Its integrated consideration of the objects allows more abstract environment e.g. by means of driving lanes. Further interpretation of a traffic situation can reduce the objects' state variables to their relevant properties and the according computational demand of the target applications. As a result a probability distribution of defined intents is assigned to the environment objects of interest.

Furthermore, the low level environment object's movement prediction estimated (s. section 4.2) by the tracking algorithm (according to the used dynamic model) can be improved by the incorporation of additional available knowledge about the intents of the objects. For this reason, a feedback of the intent estimation module to the sensor data fusion is optionally available (dashed arc in figure 4.19). Thus, maneuver level intent estimation can improve the prediction step.

The objective of this thesis concerning the intent estimation module was its incorporation into the presented integrated approach (s. chapter 4) and the proof of this concept by means of a lane change maneuver of an environmental object.

Figure 6.13 (top) shows the confidence level estimation of both the "lane keep" and the "lane change" hypotheses of the target vehicle over time by means of a selected scenario, where the target vehicle variates its lateral offset while performing a lane change maneuver. The bottom part depicts the relative lateral position of the vehicle over time. Depending on the environmental conditions and the individual driver behavior, a high probable lane change can be detected in advance before the maneuver is finished.

Due to the variability of human behavior and the complexity of traffic maneuvers a further analysis goes beyond the scope of this thesis. For a comprehensive work on this topic please refer to [SG08].

6.5 Discussion

This section summarizes the experience based on the experimental results presented in this chapter, which evaluates the main ideas introduced within this thesis. In particular, this includes the optimized track level data fusion algorithm, the novel high level self-diagnosis approach as well as the proposed intent estimation method.

The proposed sensor data fusion approach consists of three main modules: The object data fusion based on data from environmental sensors, the host vehicle movement estimation based on vehicle's built-in odometry

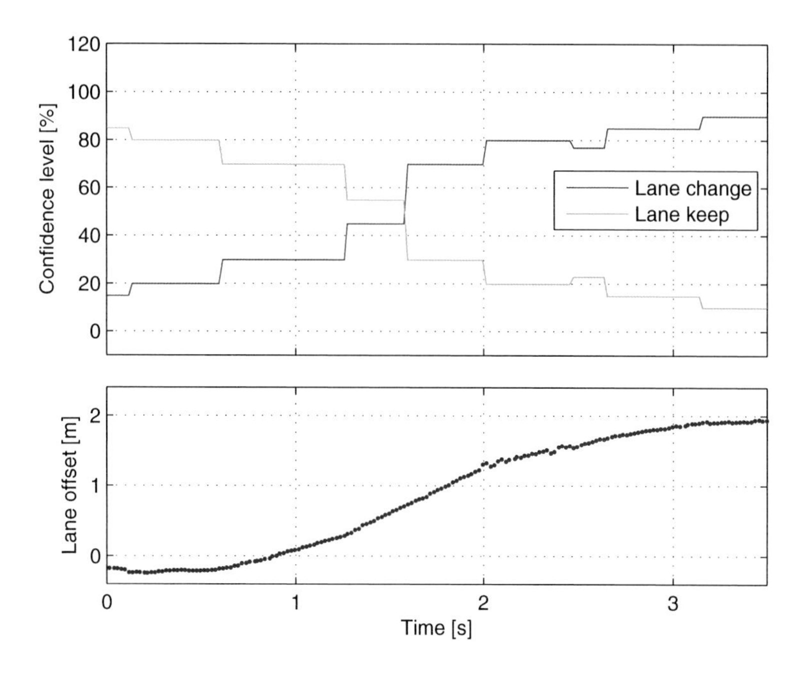

Figure 6.13: Lane change maneuver: intent estimation confidence level (top); lane offset (bottom)

sensors and the driving lane estimation based on the used camera system.

Within the object data fusion, the use of the proposed track level technique brings benefits in high dynamic scenarios (i.e. object initialization, rapid state change). Nevertheless, its successful operation depends significantly on sensors' internal (pre)tracking setup. Classical measurement level fusion produces in general more smooth results, but it can sometimes tend to overshoot and then it needs some time to settle down at a certain value. This effect is related to its higher latency.

The used host vehicle movement estimation based on on-board odometry sensors used by the ego fusion module was proved to be sufficiently precise to supply the subsequent fusion modules.

The lane data fusion module can bridge a potential short lane unavailability (e.g. due to an occlusion) successfully in order of several seconds and provides usable data for the controller application during this time.

The proposed diagnosis approach based on Bayesian networks was compared to a classical existence estimator algorithm and the standard deviation resulting from a measurement fusion algorithm. The result of the existence estimator produces quasi binary results according to the actual-theoretical comparison of sensor data. Furthermore, it strongly depends on the cycle time ratio of the used sensors. The result of the proposed diagnosis method based on Bayesian networks incorporates further influencing factors and offers additional resolution. It focuses on the cause of a potential problem, that means it can indicate before a problem (as consequence) appears.

The function of the intent estimation module is demonstrated by means of a lane change maneuver. This ability was proved to work for an arbitrary environmental object and it provides an important contribution to the presented integrated approach.

Hence the objectives of this thesis introduced in table 2.3 and theoretically derived in chapter 4 resulting in an integrated approach have been analyzed and the reached benefits as well as the limits of its individual modules are demonstrated in this chapter. In table 6.7 a summarizing rating matrix is presented.

Table 6.7 lists criteria from the addressed research areas: Data fusion, diagnosis, intent estimation and further general topics. Thereby, the classical algorithms used in the benchmarks within this chapter are complemented by an analytical calculation algorithm (e.g. TLC) and combined to one approach (left column), which is comparable to the research field of this thesis (s. 2.5). The right column of table 6.7 is in accordance with the novel integrated probabilistic approach proposed in this thesis.

The choice of an appropriate object data fusion algorithm is to be decided depending on the application and the used sensors. Basically, it will always be a trade-off between the output smoothness and the system latency. The results of a classical existence estimator reflect well the detection uncertainty of the sensors. The comparison to Bayesian network based diagnosis has shown, that an additional incorporation of the state vector and the association uncertainty can bring further benefits. An intent estimation algorithm can be incorporated within both approaches. Nevertheless, a probabilistic intent estimation assures consistency within a probabilistic approach. Although both approaches need the appropriate sensor

	Meas. fusion, existence estimator, TLC	BN track fusion, diagnosis, intent estimation
Position accuracy	+	+
Position latency	0	+
Fusion output smoothness	+	–
Diagnosis false positives	0	+
Diagnosis false negatives	0	+
Intent estimation ability	+	+
Integration level & consistency	0	+
Sensor specific part	+	–
Extensibility	0	+
Computational complexity	+	0

Table 6.7: Evaluation summary

models, the track fusion and especially the proposed diagnosis approach incorporate an additional comprehensive sensor specific knowledge. The extensibility of the probabilistic approach within a probabilistic framework is naturally given, while the extensibility of a heterogeneous approach can be limited. And finally, both algorithms are able to run in real time. Nevertheless, the computational complexity of the probabilistic approach requires slightly more resources.

Chapter 7

Summary and Outlook

A general probabilistic approach to multi-sensorial environmental perception of advanced driver assistance systems is presented. It is based on Bayesian networks, a powerful class of probabilistic models. The proposed method makes use of several modules. In particular these are the sensor data fusion module which estimates the optimal state of the environmental objects, the diagnosis module which assigns the detected objects a confidence level, and the intent estimation module which is able to compute the probability of defined maneuvers for the traffic participants. The resulting probabilities are a uniform and consistent denominator and reflect the reliability of the results. This knowledge is an important prerequisite for the development of future complex but robust advanced driver assistance systems.

The design of the used Bayesian networks represents the fundamental challenge of this approach. The causal relationships between the node variables are modeled and they define the network topology. In case of untraceable dependency among some variables, an independence assumption is to be adopted. Due to the extensible causal network topology, the network can be iteratively extended with additional nodes when an additional influence factor is identified. Besides the topology, the conditional relationships between nodes (probability tables resp. density functions) are of extreme importance. In this thesis, the principles of their estimation are referred.

Although in principle already addressed, the coupling of the on-board perception systems with digital maps can be improved in the future by on-line information. This is made possible by new networking technologies coming on the market.

Furthermore, the data fusion algorithm could be improved by multiple hypothesis probabilistic association, which could extend its application area to more complex (e.g. urban) scenarios. In this context, a further exploration of the coupling of the high level intent estimation with the low level prediction step can be done.

Finally, the confidence level estimation resulting from the diagnosis module could be further used for offline diagnosis to identify the cause of potential repetitive functional degradation.

Appendix A

Reference framework

The development of advanced driver assistance systems based on sensory environment knowledge requires a possibility to objectively evaluate the performance of sensors. If the sensor data is further processed e.g. by a filtering algorithm, the evaluation of its output is of interest as well.

Within the navigation reference process, a highly accurate pose of the host vehicle is determined by a differential global navigation satellite system (DGNSS) possibly combined with an inertial platform. Additionally, the relative position of the target object(s) is calculated. The target position is either known in advance (static case) or it is continuously measured (dynamic case). Each referenced dynamic target object must be equipped with a reference navigation system and static objects' pose must be measured before.

Today's available high-precision reference navigation systems provide in the optimal case a positional accuracy of several centimeters. By the use of each such system in the host and in the target vehicle, one can determine the target's relative position and its relative velocity and acceleration according to this precision order. The geometrical dimensions (e.g. length, width, height) of the target vehicle are known because of its a priori knowledge (to be measured in advance).

In the following a reference framework is presented, which allows both the evaluation of sensors, as well as the rating of further processing algorithms.

A.1 Hardware

Basically, different system types from various manufacturers can be used as a reference. This includes a pure differential global positioning system

(DGPS) receiver as well as an advanced navigation system with integrated inertial platform. Differences in accuracy and update rate are to be considered in the desired analysis. The relative reference pose can be determined from the absolute positions of host and target vehicles. Within this thesis the RT3003 dual antenna integrated inertial and DGPS navigation system by Oxford Technical Solutions was used. The extract of its technical parameters is listed in table A.1. The transferability to other inertial systems is provided.

Parameter	Accuracy [1σ]	
Position	0,02	m
Velocity	0,01	m/s
Acceleration	0,01	m/s^2
Roll/Pitch	0,03	°
Heading	0,10	°
Angular rate	0,01	°/s

Table A.1: Extract from manufacturer's specification of RT3003 system

In the following of selected advantages of the used reverence system over pure DGPS system are listed:

- High update rate of 100Hz.
- Low output latency of 3,9ms.
- Continuous output even during GPS blackouts (e.g. under a bridge).
- Additional measurements due to the integrated inertial platform (e.g. acceleration, angular rate, heading, pitch, roll).
- Accurate heading even in stationary case or while driving with low vehicle dynamics with constant performance.

A.2 Evaluation procedure

The evaluation method consists on the one hand of an online test with data recording and on the other hand of offline data processing and analysis. In the following an exemplary configuration test with one (dynamic) target vehicle is described.

A.2.1 Hardware configuration and test drive(s)

Initially, the host and the target vehicle are to be equipped with the reference system and the according monitoring and recording hardware (ADTF PC with protocol camera). The above described reference hardware is easily portable. Important is to measure the relative mounting point of the reference unit in both vehicles. These data are to be stored in "ini" files, one each per vehicle (once). The position of the antennas relative to the reference unit is to be set directly in this unit and various other setting can be performed depending on individual needs. Figure A.1 shows a configuration overview of the reference and recording hardware.

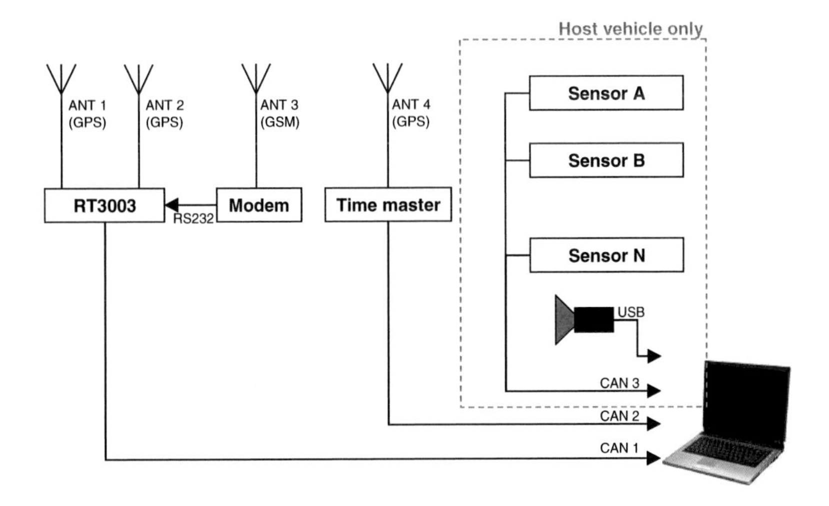

Figure A.1: Hardware configuration of the reference system

Additionally, the internal vehicle busses of both vehicles can be recorded. The evaluation of a sensor can afford the knowledge about the vehicle's driving condition (e.g. ABS or ESP intervention) to be able to consider special situations. Generally, the incorporation possibility of further vehicle data, that are typically available on vehicle bus (e.g. brake lights or turn signals) is of advantage.

The test drives can be performed manually by a driver or automated by a driving robot as described in section 4.3.2. Every test drive is to be recorded separately in the host vehicle, the recording in the target vehicle can run continuously over the entire test set. The ADTF multi-purpose ".dat" data format is suitable for all recordings. During the tests, it is important to monitor the system components status to prevent later surprises in the recorded data.

A.2.2 Offline processing

For a qualitative evaluation of some sensor or algorithm, online calculations in real time during a test drive are helpful, but an comprehensive off-line analysis should follow in any case. The aim of the reference framework tool-chain is to automate the evaluation of test runs as far as possible and to generate the results in a concise format with clear diagrams. It focuses primarily on object-detecting sensors' measurements, that are to be compared with known reference objects to provide a statement about the detection quality.

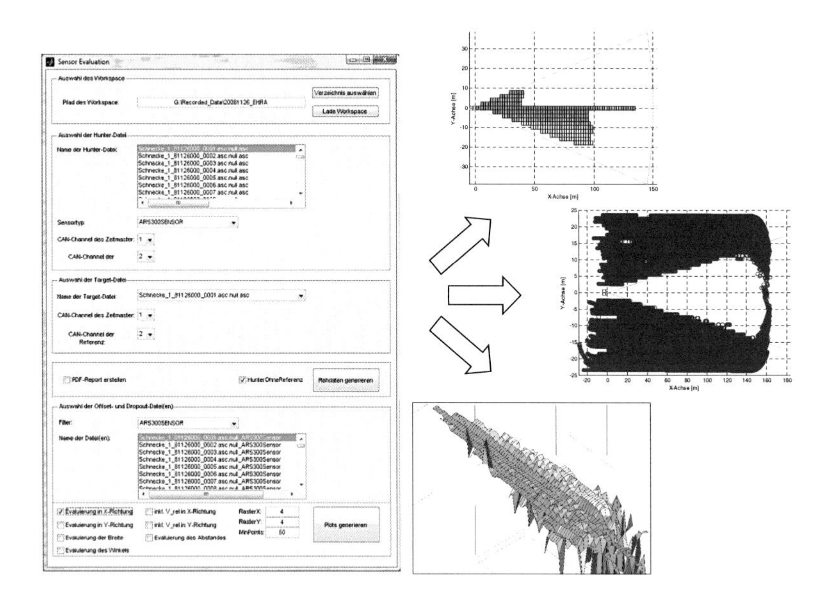

Figure A.2: Graphical user interface for offline processing

The offline processing possesses a high degree of automation. The processing tool chain can be controlled via a Matlab graphical user interface (GUI), which is depicted in figure A.2. This interface needs five files to evaluate a measurement: a "csv" file containing the object list of the sensor, the position data of the reference system each of host and target vehicle as well as the according "ini" files where the geometrical data of the vehicles and sensor positions are stored. These files can be extracted via ADTF from the recorded ".dat" files and they are to be provided per each test run.

The GUI is to be operated as follows: Initially, the working folder has to be specified. If the folder is specified, it can be read with "Load Workspace" button. The particular host and target files which are to be processed can be chosen via drop-down menu. In the background, the GUI controls a set of scripts. Thereof, the most important steps are listed in the following:

Target cut Cutting the "big" target file according to the global start and end times of the host files.

Synchronize Generation of the time synchronization file to synchronize the sensor data. This file assigns each CAN time a corresponding UTC time, which was received by the time master.

Relative reference In order to verify object-detecting sensors, the reference pose data from the sensor carrier (host) and the target are processed to a relative sensor reference, that indicates the distance between the vehicles in metric units in the host sensor coordinate system. This reference is then used for the comparison with the sensor data.

Plot generation After the preprocessing of the raw data through the upper section of the GUI, the generation of graphical results can follow. Therefore, the according data files are to be selected. The selection is simplified by a sensor filter. If the desired data files selected, you can select a checkbox, which variables are to be evaluated. Then, for each variable the evaluation results according to the reference in the entire field of view of the sensor are displayed (e.g. offsets, standard deviation, general availability of sensor measurements).

Bibliography

[AK77] H. Akashia and H. Kumamoto. Random sampling approach to state estimation in switching environments. *Automatica*, 13(4):429–434, July 1977.

[AS06] E. Ahle and D. Söffker. Entwurf komplexer Automatisierungssysteme am Beispiel eines Überwachungsautomaten für Überholvorgänge von Kraftfahrzeugen. In *Tagungsband EKA 2006*, 2006.

[Bau02] H. Bauer. *Kraftfahrtechnisches Taschenbuch*. Vieweg, 2002.

[Bla86] S. S. Blackman. *Multiple Target Tracking with Radar Applications*. Norwood, MA, 1986.

[BLS08] P. Bessire, C. Laugier, and R. Siegwart. *Probabilistic reasoning and decision making in sensory-motor systems*. 2008.

[BP99] S. Blackman and R. Populi. *Design and Analysis of Modern Tracking Systems*. Artech House, 1999.

[BS81] Y. Bar-Shalom. On the track-to-track correlation problem. *IEEE Transactions on Automatic Control*, 26(2):571–572, Apr 1981.

[BSF08] C. Blaschke, J. Schmitt, and B. Färber. Überholmanöver-Prädiktion über CAN-Bus-Daten. *Automobiltechnische Zeitschrift ATZ*, 11, November 2008.

[BSL95] Y. Bar-Shalom and X.-R. Li. *Multitarget-Multisensor Tracking: Principles and Techniques*. YBS Publishing, 3. edition, 1995.

[CFBM02] C. Coue, T. Fraichard, P. Bessiere, and E. Mazer. Multi-sensor data fusion using bayesian programming: An automotive application. In *Proceeding of the IEEE/RJS International Conference on Intelligent Robots and Systems*, volume 1, pages 141–146. IEEE - Institute of Electrical and Electronics Engineers, Inc., 2002.

[Dag05] I. Dagli. *Erkennung von Einscherer-Situationen für Abstandsregeltempomanten.* PhD thesis, Eberhard-Karls-Universität Tübingen, Fakultät für Informations- und Kognitionswissenschaften, 2005.

[DBM03] J. Diard, P. Bessière, and E. Mazer. A survey of probabilistic models, using the bayesian programming methodology as a unifying framework. In *Proc. of the Int. Conf. on Computational Intelligence, Robotics and Autonomous Systems*, Singapore (SG), December 2003.

[Dem67] A. P. Dempster. Upper and lower probabilities induced by a multivalued mapping. In *Annals of Mathematical Statistics*, volume 38, pages 325–339, Baltimore, 1967. Waverly Press.

[DKK05] K. Dietmayer, A. Kirchner, and N. Kaempchen. *Fahrer-Assistenzsysteme mit maschineller Wahrnehmung*, chapter Fusionsarchitekturen zur Umfeldwahrnehmung für zukünftige Fahrerassistenzsysteme. Springer Verlag, Berlin, 2005.

[DKP+06] P. Domingos, S. Kok, H. Poon, M. Richardson, and P. Singla. Unifying logical and statistical AI. *Proceedings of the Twenty-First National Conference on Artificial Intelligence*, pages 2–7, 2006.

[DVDD97] J. De Geeter, H. Van Bruessel, J. De Schutter, and M. Decreton. A smoothly constrained kalman filter. In *IEEE Transactions on Pattern Analysis and Machine Intelligence*, volume 19, pages 1171–1177. IEEE - Institute of Electrical and Electronics Engineers, Inc., October 1997.

[DZ86] E. Dickmanns and A. Zapp. A curvature-based scheme for improving road vehicle guidance by computer vision. *SPIE - The International Society for Optical Engineering*, 727:161–168, October 1986.

[EC98] EC. Directive 98/69/ec, Oct 1998.

[Eff09] J. Effertz. *Autonome Fahrzeugführung in urbaner Umgebung durch Kombination objekt- und kartenbasierter Umfeldmodelle.* PhD thesis, Technische Universität Carolo-Wilhelmina zu Braunschweig, Fakultät für Elektrotechnik, Informationstechnik, Physik, 2009.

[Eig09] T. Eigel. *Integrierte Längs- und Querführung von Personenkraftwagen mittels Sliding-Mode-Regelung.* PhD thesis, TU Braunschweig, 2009.

[For02] J. R. N. Forbes. *Reinforcement Learning for Autonomous Vehicles.* PhD thesis, University of California at Berkeley, 2002.

[GEJS08] T. Giebel, T. Eigel, J. Jerhot, and C. Semmler. The subproject integrated lateral assistance at half time of the german research initiative AKTIV. In *17th Aachen Colloquium "Automobile and Engine Technology"*, 2008.

[GSS93] N. J. Gordon, D. J. Salmond, and A. F. M. Smith. Novel approach to nonlinear/non-gaussian bayesian state estimation. *IEE Proceedings of Radar and Signal Processing*, 140(2):107–113, April 1993.

[Hec95] D. Heckerman. A tutorial on learning with bayesian networks. Technical Report MSR-TR-95-06, Microsoft Research, March 1995.

[HGC94] D. Heckerman, D. Geiger, and D. Chickering. Learning bayesian networks: The combination of knowledge and statistical data. In *Proc. 10th Conf. Uncertainty in Artificial Intelligence*, number MSR-TR-94-09, pages pp. 293–301, San Francisco, CA., March 1994. Morgan Kaufmann Publishers.

[HL01] D. L. Hall and J. Llinas. *Handbook of Multisensor Data Fusion.* CRC Press, June 2001.

[HM69] J. E. Handschin and D. Q. Mayne. Monte carlo techniques to estimate the conditional expectation in multi-stage non-linear filtering. *International Journal of Control*, 5(5):547–559, 1969.

[Ise06] R. Isermann. *Fahrdynamik-Regelung: Modellbildung, Fahrerassistenzsysteme, Mechatronik.* Vieweg+ Teubner, 2006.

[Jay68] E. T. Jaynes. Prior probabilities. 4(3):227–241, Sept. 1968.

[JFS$^+$09] J. Jerhot, T. Form, G. Stanek, M. Meinecke, T. Nguyen, and J. Knaup. Integrated probabilistic approach to environmental perception with self-diagnosis capability for advanced driver assistance systems. In *Information Fusion, 2009. FUSION'09. 12th International Conference on*, pages 1347–1354. IEEE, 2009.

[JMFN09] J. Jerhot, M.-M. Meinecke, T. Form, and T.-N. Nguyen. Methodik zur probabilistischen Selbstdiagnose der Umfeldwahrnehmung von Fahrerassistenzsystemen. In *Proc. of the 10. Symposium AAET 2009 - Automatisierungs-, Assistenzsysteme und eingebettete Systeme für Transportmittel*, February 2009.

[JN09] H. Jäger and D. S. Neads. Autonomes System für fahrerloses Testing. In *Proc. of the 10. Symposium AAET 2009 - Automatisierungs-, Assistenzsysteme und eingebettete Systeme für Transportmittel*, February 2009.

[JR07] J. Jerhot and B. Rössler. A sensor data fusion architecture integrated application driven simulator for evaluation of adas. In *4th International Workshop on Intelligent Transportation*, 2007.

[JU97] S. S. Julier and J. K. Uhlmann. A new extension of the kalman filter to nonlinear systems. In *Proceedings of the SPIE AeroSense Symposium*. SPIE - The International Society for Optical Engineering, April 1997.

[JU07] S. J. Julier and J. K. Uhlmann. Using covariance intersection for slam. *Robot. Auton. Syst.*, 55(1):3–20, 2007.

[Kal60] R. E. Kalman. A new approach to linear filtering and prediction problems. *Transactions on the ASME - Journal of Basic Engineering*, 82 (Series D):35–45, 1960.

[KBL+07] O. Krieger, A. Breuer, K. Lange, T. Müller, and T. Form. Wahrscheinlichkeitsbasierte Fahrzeugdiagnose auf Basis individuell generierter Prüfabläufe. In *Mechatronik 2007 - Innovative Produktentwicklung*. VDI Verlag GmbH, May 2007.

[KE08] M. Koplin and W. Elmenreich. Analysis of kalman filter based approaches for fusing out-of-sequence measurements corrupted by systematic errors. In *Proc. IEEE International Conference on Multisensor Fusion and Integration for Intelligent Systems MFI 2008*, pages 175–180, 20–22 Aug. 2008.

[KLK98] C. Kreucher, S. Lakshmanan, and K. Kluge. A driver warning system based on the LOIS lane detection algorithm. In *Intelligent Vehicles Symposium*, Stuttgart, November 1998.

[Kop00] S. Kopischke. *Entwicklung Einer Automatischen Notbremsfunktion mit Rapid Prototyping Methoden*. PhD thesis, Technische Universität Braunschweig, Fachbereich für Maschinenbau und Elektrotechnik, February 2000.

[Lan01] J. Langheim. CARSENSE - new environment sensing for advanced driver assistance systems. In *Proceedings of the IEEE Intelligent Vehicles Symposium*, Korea, 2001. IEEE - Institute of Electrical and Electronics Engineers, Inc.

[Löb08] C. Löbel. ADTF: Framework für Fahrerassistenz- und Sicherheitssysteme. In *Proceedings of Fisita 2008*, 2008.

[MA08] S. Matzka and R. Altendorfer. A comparison of track-to-track fusion algorithms for automotive sensor fusion. In *IEEE International Conference on Multisensor Fusion and Integration for Intelligent Systems*, 2008.

[Mah36] P. C. Mahalanobis. On the generalised distance in statistics. In *Proceedings National Institute of Science, India*, volume 2, pages 49–55, April 1936.

[Mau00] M. Maurer. *Flexible Automatisierung Von Straßenfahrzeugen mit Rechnersehen*. PhD thesis, Universität der Bundeswehr München, 2000.

[May79] P. S. Maybeck. *Stochastic models, estimation, and control*, volume 141 of *Mathematics in Science and Engineering*. 1979.

[ME84] H. P. Moravec and A. Elfes. High resolution maps from wide angle sonar. Technical report, Carnegie-Mellon University, The Robotics Institute, 1984.

[MM96] M. C. Martin and H. P. Moravec. Robot evidence grids. Technical report, Carnegie Mellon University, The Robotics Institute, 1996.

[MP43] W. McCulloch and W. Pitts. A logical calculus of the ideas immanent in nervous activity. *Bulletin of Mathematical Biophysic*, (5):115–133, 1943.

[MRD06] M. Maehlisch, W. Ritter, and K. C. J. Dietmayer. ACC vehicle tracking with joint multisensor multitarget filtering of state and existence. *PReVENT Fusion Forum e-Journal*, 1:37–43, 2006.

[Mur02] K. P. Murphy. *Dynamic Bayesian Networks: Representation, Interference and Learning.* PhD thesis, University of California, Berkeley, 2002.

[MW01] M. Moore and J. Wang. Adaptive dynamic modelling for kinematic positioning. 2001.

[NMM+08] T.-N. Nguyen, B. Michaelis, M.-M. Meinecke, T.-B. To, and J. Jerhot. A sensor fusion approach based on occupancy grid and fuzzy logic. In *5th International Workshop on Intelligent Transportation*, 2008.

[NMM+09] T.-N. Nguyen, B. Michaelis, M.-M. Meinecke, J. Jerhot, and J. Knaup. Ein Kombinationsansatz kartenbasierter und objektbasierter Umfeldmodellierung für Fahrerassistenzsysteme. In *10. Braunschweiger Symposium AAET 2009 - Automatisierungs-, Assistenzsysteme und eingebettete Systeme für Transportmittel*, 2009.

[OP00] N. Oliver and A. P. Pentland. Graphical models for driver behavior recognition in a SmartCar. In *Intelligent Vehicles 2000*, 2000.

[Pap01] L. Papula. *Mathematik für Ingenieure und Naturwissenschaftler.* Vieweg, 4 edition, 2001.

[Pea88] J. Pearl. *Probabilistic reasoning in intelligent systems: networks of plausible inference.* Morgan Kaufmann, 1988.

[Rüd03] J. Rüdenauer. Einsatz probabilistischer Verfahren zur Entscheidungsfindung im RoboCup. Master's thesis, Universität Stuttgart, Institut für Parallele und Verteilte Systeme, 2003.

[Rec73] I. Rechenberg. *Evolutionsstrategie: Optimierung technischer Systeme nach Prinzipien der biologischen Evolution.* Problemata. Frommann-Holzboog, Stuttgart-Bad Cannstatt, 1973.

[RM01] H. Rohling and M. M. Meinecke. Waveform design principles for automotive radar systems. In *Proc. Radar CIE International Conference on*, pages 1–4, 15–18 Oct. 2001.

[RME00] R. Risack, N. Möhler, and W. Enkelmann. A video-based lane keeping assistant. In *Intelligent Vehicles Symposium*, 2000.

[RN03] S. Russell and P. Norvig. *Artificial Intelligence: A Modern Approach (2nd Edition)*. Prentice Hall, 2003.

[Ros57] F. Rosenblatt. The perceptron: A perceiving and recognizing automaton. Technical Report 85-460-1, Cornell Aeronautical Laboratory, Ithaca, NY, 1957.

[RW99] S. Renooij and C. Witteman. Talking probabilities: communicating probabilistic information with words and numbers. *International Journal of Approximate Reasoning*, 22(3):169–194, 1999.

[SG08] F. Schroven and T. Giebel. Fahrerintentionserkennung für Fahrerassistenzsysteme. In *VDI-Berichte, Band 2048: Proceedings der 24. VDI/VW-Gemeinschaftstagung - Integrierte Sicherheit und Fahrerassistenzsysteme*, 2008.

[Sha76] G. Shafer. *A mathematical theory of evidence*. Princeton University Press, Princeton, USA, 1976.

[SK71] R. Singer and A. Kanyuck. Computer control of multiple site track correlation. *Automatica*, 7(4):455–463, Jul 1971.

[SMI+08] M. Schulze, T. Mäkinen, J. Irion, M. Flament, and T. Kessel. Ip prevent final report. Technical report, PReVENT Consortium, 2008.

[SMKY07] D. D. Salvucci, H. M. D. Mandalia, N. Kuge, and T. Yamamura. Lane-change detection using a computational driver model. *HUMAN FACTORS*, 49:532–542, 2007.

[Stü04] D. Stüker. *Heterogene Sensordatenfusion zur robusten Objektverfolgung im automobilen Straßenverkehr*. PhD thesis, Carl von Ossietzky-Universität Oldenburg, September 2004.

[TBF05] S. Thrun, W. Burgard, and D. Fox. *Probabilistic Robotics*. The MIT Press, 2005.

[TGW+05] K. Torkkola, M. Gardner, C. Wood, C. Schreiner, N. Massey, B. Leivian, J. Summers, and S. Venkatesan. Toward modeling and classification of naturalistic driving. In *Intelligent Vehicles Symposium*, Las Vegas, USA, June 2005.

[Thr02] S. Thrun. Particle filters in robotics. In *Proceedings of the 17th Annual Conference on Uncertainty in AI (UAI)*, volume 1, 2002.

[TMS+08] T. B. To, M.-M. Meinecke, F. Schroven, S. Nedevschi, and J. C. Knaup. CityACC - on the way towards an intelligent autonomous driving. *IFAC 2008*, 2008.

[Uhl03] J. Uhlmann. Covariance consistency methods for fault-tolerant distributed data fusion. *Information Fusion*, 4(3):201–215, September 2003.

[WB95] G. Welch and G. Bishop. An Introduction to the Kalman Filter. Technical report, University of North Carolina at Chapel Hill, Dep. of Computer Science, 1995.

[Wei07] K. Weiß. *Interpretation von Fahrumgebungen für Fahrerassistenzsysteme.* PhD thesis, Universität Rostock, 2007.

[WSK03] K. Weiß, D. Stüker, and A. Kirchner. Target modelling and dynamic classification for adaptive sensor data fusion. In *Proceedings of the IEEE Intelligent Vehicles Symposium*, Ohio, June 2003. IEEE - Institute of Electrical and Electronics Engineers, Inc.

[XxHS07] H. Xiao-xuan, W. Hui, and W. Shuo. Using expert's knowledge to build bayesian networks. In *Proc. International Conference on Computational Intelligence and Security Workshops CISW 2007*, pages 220–223, 15–19 Dec. 2007.

[Zad65] L. A. Zadeh. Fuzzy sets. *Information and Control*, 8:338–353, 1965.

[Zad94] L. A. Zadeh. Soft computing and fuzzy logic. *IEEE Software*, 11(6):48–56, Nov 1994.